The Jigsaw of Life

The Jigsaw of Life

Richard Bryant-Jefferies

iUniverse, Inc.
New York Lincoln Shanghai

The Jigsaw of Life

iUniverse books may be ordered through booksellers or by contacting:

iUniverse
2021 Pine Lake Road, Suite 100
Lincoln, NE 68512
www.iuniverse.com
1-800-Authors (1-800-288-4677)

ISBN: 978-0-595-48002-9 (pbk)
ISBN: 978-0-595-60105-9 (ebk)

Printed in the United States of America

That is what it is to be human.
To make yourself more than you.[1]

From space I saw the Earth—
indescribably beautiful
with the scars of national boundaries gone.[2]

My country is the world, and my religion is to do good.[3]

---

1. Captain Jean-Luc Picard, speaking to Commander Data, Star Trek 2nd Generation, in the film *Nemesis*.
2. Muhammad Ahmad Faris, Syrian cosmonaut. Quoted in *The Home Planet*. KW Kelley (Ed) Addison Wesley Publishing Company. (1998)
3. Thomas Paine, Rights of Man, 1791

Contents

Foreword—Steve Nation

'I don't believe in gurus', the artist Frederick Franck told an interviewer for *Tricycle* magazine shortly before his death. 'I believe that each one of us has a riddle to solve, the riddle of what it means to be human. When we are born we are some kind of hominid, a little anthropoid animal with the potential and the capacity to become human.'

It is this potential to become human, truly, fully human, that is the subject of Richard Bryant-Jefferies book. *The Jigsaw of Life* draws on the author's extensive experiences in the Carl Rogers inspired Person Centred Approach to counselling and therapy together with his many years of studies in metaphysical and esoteric philosophies. Through a series of gentle reflections from a wide variety of perspectives the reader is drawn to consider what it means to be a human being.

The process of becoming fully human is likened here to a jigsaw, something of a puzzle as we seek to bring together disparate parts and to make sense of key experiences. A metaphysical and esoteric approach suggests that the puzzle can only really be seen whole (the completed picture of the jigsaw) when we open ourselves to the perspective of altitude. This is a view of Self held in the ancient (and ageless) traditions of all human cultures. In order to understand our potential for fullness as individuals we need to see ourselves in relation to our sublime core—in relation to the heights of our Selfhood. Repression of the sublime is as much, if not more, a cause of alienation and disorientation as is the more widely recognised suppression of the shadow. In a time when integration, synthesis, wholeness and unity are increasingly driving the evolutionary thrust (in culture, art, psychology, religion and science) it is the balance, mediation and alignment between universal and personal aspects of self that holds the key to finding and expressing our true individuality.

Making this alignment process conscious, dynamic and pro-active is increasingly becoming a keynote of the psychology, metaphysics and spirituality of our time. It calls on us to seek out those ideas and practices which lead us into relationship with the Higher Self, Soul or Buddha nature. One of the great joys of the age is that wisdom traditions are now widely available and accessible to all. At a time when personal/social/global problems caused by our competitive separateness threaten the health and well-being of our selves and our communities it matters greatly that more and more of us are exploring the ideas and practices of these wisdom traditions. It is through this deep work that cultures of interdependence are in process of forming—and that the fire of love is re-emerging as a potent force in human affairs.

Alongside the dominant schools of wisdom traditions which guide the integrative/synthesis movement (the variety of approaches to Buddhism, Christian contemplative traditions, cabbalistic approaches, indigenous pathways, transpersonal psychology, dream-work and so on) there are a number of teachings on an expanded mind and heart that are not so much in the spotlight. While remaining true to the practices and tradition which have shaped our quest it can be useful to be open to insights from some of these lesser known pathways.

In his reflections on the *Jigsaw of Life*, Richard Bryant-Jefferies draws on one of these lesser known streams—the teachings contained in a series of books by Alice A. Bailey. One of the particular points of focus that is worthy of attention in these books is the nature of soul, the essential self with its core elements of love-wisdom, inclusive intelligence and spiritual will (or sense of purpose). This central identity is seen to be universal in its outlook; naturally selfless and concerned with the good of the whole. The dynamic of the relationship between soul and personality, particularly at the stage when the personality becomes aware of its sublime core, as well as the contemplative/meditative practices that draw on the intuitive insight of the soul are major themes in the psychological approach developed by Bailey. Service as an expression of the loving mind and the intelligent heart is seen to be a natural result of all work with the higher self—regardless of the particular teachings or tradition we draw from.

While Alice Bailey was writing in the first half of the twentieth century the teachings on service speak with a direct voice to the core issues and dilemmas of these first years of a new millennium. Discipleship is another theme that is explored in a real and living way—this is discipleship in service of the evolutionary momentum towards cultures of synthesis and increasing levels of cooperation and wholeness. It is a not unusual experience for us to find today that our sense of meaning and purpose revolve around the contribution we can make (perhaps in small, personal, appropriate ways) to the emergence of an entirely new culture of right (or should one say better, more harmonious) relations. The key challenge, or so it seems to me, is for those inspired by the inclusive vision of what the United Nations describes as 'Cultures of Peace' to consciously train themselves to serve that vision with increasing clarity, intelligence, wisdom and heart—in other words to walk the discipleship path.

Another major theme that strikes me as being of particular relevance to the post-modern world is the Bailey writings on group consciousness. In the world of integrative studies and of a secular spirituality there is widespread interest in the nature of group consciousness and group intuition. The literature was well summarised in the May 2004 issue of *What is Enlightenment* on the theme of Collective Intelligence—and there is a wealth of material on the web at: http://www.wie.org/collective/. Bailey sees the awakening of a sense of group (in which the individual chooses freely and without any repression of Self or negative group dynamics to identify themselves as part of a group field, a group consciousness) and particularly of groups oriented towards service to the whole, as a natural human response to soul consciousness—and an expected evolutionary development. A number of the suggested meditative practices draw on this sense of group with visualisations and ponderings centred around identification with (immersion in) the collective mind/heart of all human beings who love (all whose lives express goodwill); imagining this centre of group consciousness overshadowed by the great potencies of compassion recognised down the ages by all the religions; invoking the universal qualities of divinity; visualising the irradiation of human consciousness with light and love and the power of enlightened purpose. This work of world service through the potencies of mind and heart is not seen in any way separate or apart from, the way we live our life, or the way we approach issues of relationship or of personal difficulty, or the way in which we are engaged in the community. We live in a time when the fusing, or to be more accurate, the conscious relating and ultimately harmonising, of inner and outer, 'higher' and 'lower' are central to health and well-being. Insights from the Bailey teachings have much to offer the deep thinker to help facilitate this process.

It is my hope that in reading this book by Richard Bryant-Jefferies, some will be led through their reflections on the jigsaw of life to dip into the 'blue books' by Alice Bailey—in addition to all the wonderful teachings now available.

Steve Nation
Co-founder and focaliser of Intuition in Service
and the United Nations Days and Years Meditation Initiative (www.intuition-in-service.org), writer and actively involved with the Darjeeling Goodwill Animal Centre in India and the Spiritual Causus at the United Nations, New York.

Foreword—Irene Fairhurst

When Richard asked me to do this foreword for his book, he said he wanted it from a Person-Centred perspective, so this is what I shall try to do.

I first met Richard over 15 years ago on his training programme in the Person-Centred Approach. He embraced the values and concepts of the work whole-heartedly, both as a therapist and as 'a way of being', and these values and concepts are inherent as themes throughout the book as Richard endeavours to answer such searching questions as 'What is life all about?', 'What does it mean to be human?', 'What do we mean by spirituality?'.

I found the book challenging in introducing new ideas, which I shall look at later, and informative, including references to so many different philosophers and writers in the fields.

The book will be meaningful for practitioners in, or people committed to, the Person-Centred Approach and who are interested in a spiritual dimension in their work or lives. However, it will also provide important reading for those people who are searching for their own answers in trying to make sense of these type of questions and have no prior knowledge or interest in the Person-Centred Approach, as Richard, as usual, gives detailed theoretical background to his ideas and conclusions.

One of these themes is that of the Actualising or Formative Tendency, Rogers' theory of personality, which can be found detailed on the first page of *Chapter 12 Person-Centred Psychology....* but which is hinted at as early as the last page of *Chapter 1, The Cosmic Jigsaw*:

'We have a long way to go, or so it seems, when we look at the world today, but I would anticipate that harmony and beauty will one day become reflected in the picture as it finally takes shape and its secret is revealed. What is this secret? What is the final picture? We do not know. For us, as pieces groping our way forward in our own small worlds of experience it remains a mystery. Yet we continue to experience this urge to grow, to find our place, to move towards realising some unknown potential, something deep within us that yearns to be known and understood.'

In looking at the world today, Richard shares a sorrow around materialism and separatism and hopes that in completing the jigsaw these world values will be replaced by values such as compassion and unanimity. This reminds me of Rogers' writing in 'Carl Rogers, the Man and His Ideas', edited by Richard I. Evans[1] where Rogers lists qualities which he has identified in 'The Emerging Person' including:

The Unimportance of Material Things
A Non-moralistic Caring
The Wish for Intimacy
The Universe Within

Knowing Richard and his passionate commitment to equality and diversity, it does not surprise me that he devotes an entire chapter to this subject: Chapter 6, *Seeking Equality in a Diverse World.* In this chapter he asks the question 'What is belief?' and warns about the dangers of having a fixed belief and '.... we see people desperately trying to make a certainty out of a belief.' This also takes me back to *Rogers and 'The Emerging Person'*, where he asks whether this is a viable person and suggests that he will meet with opposition from six main sources—the sixth being

'Our truth is *the* truth—The true believer is also the enemy of change, and he will be found on the left, on the right, and in the middle. He will not be able to tolerate a searching, uncertain, gentle person ... he must oppose this process individual who *searches* for truth. Such true believers *possess* the truth, and others must agree. So, as this person of tomorrow continues to emerge into the light, he will find increasing resistance and hostility from these six important sources. They may very well overwhelm him.'

As mentioned earlier, in *Chapter 12, Person-centred Psychology: From the Personal ...* Richard outlines Person-Centred personality theory, describing and exploring the concept of the actualising or formative tendency and how it is blocked by 'negative conditioning' and then manifested in Client-Centred therapy. In this chapter Richard introduces a new concept which he calls 'false-conditioning'—a conditioning in addition to the 'negative conditioning' postulated by

1. Evans, Richard I (1975) Carl Rogers: The Man and His Ideas. Dutton, New York, pp. 165/8

Rogers, where persons are subjected to conditioning 'that are at odds with our inner soul natures'.

In *Chapter 13 Person-centred Psychology ... to the Transpersonal*, Richard puts forward ideas that many will find challenging in that he is suggesting that Person-Centred therapy could embrace a more spiritual or soul-centred dimension. He feels that many therapy training programmes neglect this aspect of Rogers' teachings and quotes him:

'Our experiences in therapy and in groups, it is clear, involves the transcendent, the indescribable, the spiritual. I am compelled to believe that I, like many others, have underestimated the importance of this mystical, spiritual dimension'.[2]

In *Chapter 14 The Jigsaw of Self*, Richard includes a section on Client-Centred therapy in which he talks about the experiencing of the client and therapist in relationship, which contributes to building the Jigsaw of Life.

Throughout this book, Richard has shared many of his stimulating ideas and findings, and he has also disclosed much of himself and his personal beliefs—a courageous undertaking. It is impossible in a foreword (although somewhat lengthy) to do justice to the breadth and depth of this book—I hope you will be enticed to read further.

Irene Fairhurst
Client-centred counsellor/psychotherapist
Person-centred supervisor, consultant and trainer
Co-founder, British Association for the Person-Centred Approach
and Institute for Person Centred-Learning

2. Rogers, C (1980) A Way of Being. Houghton Mifflin Co. Boston p.130

About the author

Richard Bryant-Jefferies has been interested in spirituality and a quest to understand what life is all about for most of his life. From reading Eastern Mysticism in his early teenage years, he found a series of author's that took him on his own journey of spiritual understanding. Some of the names that follow may be known to you: Yogi Ramacharaka, Joel Goldsmith, Krishnamurti, Vera Stanley Alder. In particular, the ideas expressed by Alice Bailey, a spiritual philosopher writing in the first half of the 20th century had particular meaning. A growing interest in psychology and group processes took him into another phase of reading and learning, initially encountering the ideas of Carl Jung, but later finding himself more drawn to the relational and human emphasis of Carl Rogers.

By then Richard was training as a counsellor, specifically in the principles and practice of person-or client-centred therapy, the approach founded by Carl Rogers. What was particularly important was finding an approach to therapy that seemed to have the potential to embrace Richard's ideas and understanding of the nature of spirituality, and in particular the ideas that human beings had greater potential than they often realised. Also, that there was a universal impulse towards growth or development.

It was after training as a counsellor that Richard began working in the field of addiction, primarily working with people who had alcohol problems. At around the same time he developed with his then partner, Lynn Frances, a series of workshops which used sacred circle dance and visualisation to explore the human personality and the spiritual dimension. They co-authored *The Sevenfold Circle: Self Awareness in Dance (1997)* in which Richard included some of his ideas in relation to spirituality and psychology, the 'sevenfold' in the title referring to a model of human personality based on seven fundamental types, an idea drawn from the writings of Alice Bailey.

As his involvement in working with problem drinkers developed, Richard began writing articles, running workshops and giving presentations at Conferences. Towards the end of the 1990s he began to put down on paper his ideas for working with people who had alcohol problems from a person-centred perspective. The result was *Counselling the Person Beyond the Alcohol Problem (2001).* Whilst this book drew together many different strands of thinking on this topic, the aspect of this book that proved to be of greatest significance for Richard were the fictitious dialogues written to illustrate counselling processes.

In 2001 Richard began to write a new book which was to become *Problem Drinking: Person-Centred Dialogues (2003).* This was a set of fictitious counselling and supervision sessions taking a client through the process of resolving an alcohol problem. It was written in an almost a novel-like format but with theoretical comment boxes within the text and questions for discussion at the end of each chapter. It was a refreshing approach to conveying the process of counselling and psychotherapy, and in particular from the person-centred perspective.

There then followed a period of intense writing, and to this day Richard finds it hard to make sense of how much was written over such a short space of time. *Problem Drinking* was published in January 2003. By the end of that year, five more titles in what became the *Living Therapy* series had also been published by Radcliffe Publishing. By the end of 2006 Richard had written and had published a total of 17 *Living Therapy* titles, each in a similar format to *Problem Drinking.* Topics addressed included: time-limited therapy in primary care, counselling an adult survivor of child sexual abuse, counselling a recovering drug user, young people, progressive disability, sons and mothers, supervision, mental illness, obesity, workplace issues, eating disorders in men and in women, problem gambling, victims of warfare, young binge drinkers, and death and dying. He also wrote *Models of Care for Drug Service Provision* during that period. Thirteen of the *Living Therapy* titles are being published in Chinese in China.

In 2003 Richard took on a new job, as the manager of NHS substance misuse services in Kensington and Chelsea, London. By this time Richard had met Movena Lucas, his partner, who was instrumental in encouraging him both to apply for that job and to continue to write the Living Therapy titles. Her own background as a nurse working with people with mental health and substance misuse issues, and her particular skill and experience of working with young people, contributed to helping Richard's writing in these areas.

Richard had also been aware of how often clients did not make it to counselling sessions and he wanted to write something for them to encourage those who are facing despair in their lives. Though written a few years previously, 2006 saw Richard self-publish *A Little Book of Therapy*, a dip in and out of book comprising statements of despair followed by empathic responses providing ideas and direction.

Many reviewers have commented on how Richard's *Living Therapy* titles read like a 'can't-put-it-down novel'. As a result, Richard decided to write his first novel, entitled *Binge!,* published in 2007 by iUniverse. It is centred around the experience of one man in adulthood trying to resolve his alcohol use which stems from childhood trauma, and the role of counselling/psychology in that process. It is planned that this will be the first in a series of novels, each designed to explore a contemporary emotional landscape in the context of therapy. He is currently working on a new novel, with the working title of *Alive and Cutting*, which addresses teenage self-harm in a therapeutic context.

Today Richard is an Equalities and Diversity Lead within an NHS Trust that provides mental health, substance misuse and a range of other specialist services within London. The relational component remains a theme for him once more as Equalities and Diversity is very much about how people relate to each other. He also continues to offer a counselling supervision practice in Surrey and workshops on a range of themes. Further information about this and his books can be found at: http://www.bryant-jefferies.freeserve.co.uk

Introduction

What is life all about? Where is it taking us? What does it mean to be a human being? What do we mean by 'spirituality'? How can we put all our experiences together in some kind of meaningful way? Is life some chance occurrence on one large rock spinning around an insignificant sun on the edge of the universe? Or is there a larger picture to life, a greater unseen purpose at work and, if so, what is it?

Over the years I have found myself reflecting more and more on these kinds of themes and as a result writing a number of short articles. More recently, I have begun to review these writings and the idea came to mind of putting them together as a book. For me, it is a process of assembling my own jigsaw, bringing together ideas and putting them together so that I can see a larger picture of myself as a human being and of my role and place within the larger scheme of things. It will not be the kind of book in which each chapter always flows to the next, though some might. Rather it will be one in which each chapter stands on its own, though there are going to be threads of connection. I envision this book as a kind of lucky-dip of ideas that I hope you, the reader, will find thought-provoking. I shall also intersperse some poetry, offering another path to connecting with deeper matters.

Some of the ideas that I present will lead to conclusions, others will be left awaiting further insight. Perhaps I will discover these answers later in my own life, or maybe you, the reader, will have your own answers and will feel you have developed a theme further than I have.

You may agree with what I write, you may disagree, I really do not mind. Most of what I write is my own attempt to make sense of my experiences. It will be drawn from many sources and some of it will overlap with what others think, some of it will not. As I start out writing this book I do not know what it will look like in its final form. That feels to me rather like a jigsaw when you do not look at the picture on the cover of the box. I am surrounded by pieces and I am not yet sure exactly how they are going to come together, and what it will look like, or indeed whether I will finish it completely. And I am sure that there will be some pieces missing.

So I invite you to share with me in this journey of putting together ideas and wonderings and maybe, by the end of the book, we will both have a clearer sense of ourselves, our place in the jigsaw of life and a few images of what the greater picture may look like. And perhaps as well as some answers, we will each have further questions to pursue as we seek to make sense of what I am here terming 'The Jigsaw of Life'.

The Breath of Life

The breath of life within us dwells;
The life that's mine is yours as well.
Yet worlds apart, the gaps appear;
Distinction rules, self-conscious here.
Connections hidden, veiled from sight
Exist outside of human light.
But still the breath, our common ground;
Rhythmic pulse without a sound.

Through the veils the surging breath.
No time, no space, no life, no death.
It fills the void, we glimpse our goal,
The gaps are gone, we touch the Whole.
Driven by an unknown Will
The tension holds, in silence, still.
The breath of Life now flowing free
Unites the One Humanity.

1

The Cosmic Jigsaw[1]

We exist in the world as apparent fragments of a divided whole; pieces, perhaps, of some vast, cosmic jigsaw. What is the task before us? What is our role, our place within this great jigsaw of life? Perhaps it is to create a picture, the picture that is held within some far greater Universal Mind—perhaps the Mind of God? We are in some way pieces of the final picture. Or perhaps we should say pieces that have the potential to in some mysterious way be, or become, that final picture. Yet we start out not as separate pieces moulded in shape for the final assembly, but rather more like plain pieces of card, having neither shape, colour or pattern. It seems that it is left for us to find our true shape and to mould ourselves to it so that it conforms to that which is envisioned within the Mind of God. The same is then true for colour and pattern. We begin with neither and are destined to tread an evolutionary path so that we might discover what they are, build them into ourselves and reflect them through the manner of our daily living. It seems that we are being asked to consciously discover for ourselves our true nature, our inner potential as human beings, and to give this living expression.

However, even when we have created some measure of order within our own piece of the jigsaw of life, we have still to find those other pieces to which we must be joined. We are aided by our achievements thus far; we are making progress and the creation of an integrated form involving shape, colour and pattern that is our own piece—ourselves—enables us to begin to sense or glimpse something of the larger pattern that is the greater picture of which we are a part.

Having achieved integrity within our own piece we may discover that we are rather like pieces of a shattered holographic plate, containing within ourselves a blurred image of the whole. Shine coherent light upon such a fragment and the whole image of which it is a part will appear, though in a less distinct form to

1. The chapter is an updated version of an article by the author and first published in The Beacon (1984) Vol. 1 No 11. Lucis Press Ltd, London and New York.

what it would be if the holographic plate was complete. As we realise our full potential we realise, too, that we are linked in some deep and intimate way with creation. We, the part, somehow also contain within us the whole. We gain a vision of wholeness, of unity and of an essential oneness. It provides us with a living link through which we gain a fuller sense of the greater picture. It can offer us the possibility of gaining a clearer realisation of the part that we play in building the bigger picture in our world, of the qualities and the values that we need to bring into the circle of our daily lives to contribute to the greater process of humankind forming itself into an image that reflects the Mind or Purpose of God, what we might term as His 'Plan for humanity'.

In so doing, maybe we also encourage others to seek out their potential, to find the shape, colour and pattern that is the potential for their piece in the cosmic jigsaw. These coloured patterns might be regarded as the qualities we evolve within our physical, emotional and mental natures. Our own process may in some mysterious way stimulate the process in others. Might it not be that as we as individuals seek to more consciously evolve and to find our place within the greater picture, others seeing us will begin to wonder about themselves and their part in the overall picture? They may awaken to their potential as being more than the pieces of card with shape, colour and pattern that simply satisfy their immediate needs. They may begin to experience an urge or the sense of a need to reshape themselves to fit into the bigger picture. They may wish to explore what colours, shapes and patterns they too might become.

Also, we may begin to attract to us those with whom we are to forge a link, other pieces feeling akin to us by virtue of their own shapes, colours and patterns, who hold to similar values and principles as ourselves. Perhaps the pieces that are us are to some degree magnetic, but not in the sense of attracting any piece that is close, but only those that have some kind of commonality, some kind of mutual resonance.

We may be tempted to ask who is actually doing the jigsaw, who is it that is putting the pieces together as they discover their essence, shape, colour and pattern? Is it God, or some other Divine Agency at Work? Or is it a vast process that, under law, has been set in motion and moves inexorably to an inevitable completion? Certainly, the whole process is governed by law and it is the sequence of causes set in motion under law that governs the process. Are these laws purely the physical laws sought by scientists and mathematicians? Or are there other laws that remain undiscovered governing aspects of our nature that are beyond the known and the currently measurable?

Could it simply be that it is us, humanity, that is doing the jigsaw, that we are in some way both the pieces and the builders, self-directed in our creativity yet governed by cosmic laws? Whilst we may work under the guidance and limitations of evolutionary forces, are we not free as well to choose our own ways towards achieving the final picture? Is the final picture predestined? A goal is inevitable, but what will it be? Can it be known beforehand? Are we limited by that which is envisioned by the Divine, or which is governed by Law? Perhaps the goal is inevitable, but free will lies with our choosing, our freedom to choose how soon and by what path we will reach it.

It is difficult to visualise the myriad of energies involved in the evolutionary process, and to recognise the vast number of relationships that exist and are a vital part of the process. It seems too immense, too all-encompassing for our minds to grasp. But metaphor can help us. We can visualise a jigsaw. We can imagine pieces gradually moulding themselves into shape. We can see them developing colour and pattern, perhaps as a result of contact with other pieces and their colours and patterns, or in response to glimpses of the greater picture, of a Divine Pattern.

We can imagine them grouping together, each attracted to those that have some similarity. Each piece remains unique, yet they group together not because of sameness, but because they contain common factors—perhaps a line on one edge of a particular shade of colour provides the match, the basis for a relationship. Does the shape match as well? Maybe, maybe not. Perhaps the pieces will evolve into a partnership, perhaps they will move away, destined for some other connection.

In time, complementary pieces find each other, and find the group of pieces with which they belong. We co-operate through our contact to create clusters that contain pieces of the final image. As we come into group relationship and a sense of group consciousness we are able to sense a little more clearly the design of the greater picture. Such groups may form, for instance, to take forward a particular service initiative in the world, or to focus scientific thought to deepen human understanding. Or they could be groups of friends or work colleagues. We may then, through greater co-operation, need to move and evolve with the group, and attract other pieces destined to share in our part of the picture.

Perhaps we are each more than just one piece. Could we be creating our own internal jigsaw too, drawing together apparent opposites into complementary relationships in order to create our own inner picture, our own microcosmic jigsaw so that our piece may evolve and finds its place in the greater, cosmic jigsaw?

Slowly, we find our groups and gradually larger patterns emerge. Nothing, though, is static. It can seem as though the pieces have to evolve through many permutations as part of their own process of finding that final shape, colour and pattern. Colours blend and match. Intricate shapes find their partners and are joined and yet, as the overall picture begins to take shape, the form of each individual begins to lose its significance. The colours and patterns become far more important once the right connections are made and established. The sense of separation, of being a separate piece in the jigsaw fades away. Can we also maintain our individuality? Or does this, too, blend into the groups of pieces with identity taking on a whole new meaning? Perhaps we identify then with the group, with the part of the final picture of which we are a part, or perhaps, like the piece of the holographic plate, we begin to sense the presence of the greater picture within us. We are no longer alone, individual and isolated. We become one with the greater process, whether we call that the working out of God's Purpose or the inexorable effect of evolutionary movement under scientific law, or both.

We have a long way to go, or so it seems when we look at the world today, but I would anticipate that harmony and beauty will one day become reflected in the picture as it finally takes shape and its secret is revealed. What is this secret? What is the final picture? We do not know. For us, as pieces groping our way forward in our own small worlds of experience it remains a mystery. Yet we continue to experience this urge to grow, to find our place, to move towards realising some unknown potential, something deep within us that yearns to be known and understood.

Perhaps someone is observing the process. Perhaps as harmony and beauty become revealed an Onlooker will begin to smile. Perhaps when this occurs it will be drawing near to the end of the seventh day of creation and the Onlooker will see that it is Good and that it is suited to the Purpose that He has in Mind.

◆ ◆ ◆

Instead of an intellectual search, there was suddenly a very deep gut feeling that something was different. It occurred when looking at Earth and seeing this blue-and-white planet floating there, and knowing it was orbiting the Sun, seeing that Sun, seeing it set in the background of the very deep black and velvety cosmos, seeing—rather, knowing for sure—that there was purposefulness of flow, of energy, of time, of space in the cosmos—that it was beyond man's rational ability to understand, that suddenly there was a nonrational way of understanding
that had been beyond my previous experience.
There seems to be more to the universe than random, chaotic, purposeless movement of a collection of molecular particles.

On the return trip home, gazing through 240,000 miles
of space toward the stars and the planet from which I had come,
I suddenly experienced the universe as intelligent, loving, harmonious.[2]

2. Edgar Mitchell, US astronaut. Quoted in The Home Planet. KW Kelley (Ed) Addison Wesley Publishing Company. (1998)

2

A New Era

We live during a period of great change; a time of new ideas and shifting values. Many write and speak of a 'New Age' coming upon us and I wonder sometimes just what is meant by this. There is often much confusion as to just what 'New Age' means. Some feel threatened by it; some believe it to be the answer to everything. My guess is that we will all have our own perception based on our experiences and beliefs. I would like to share my own views, not with the intention of trying to make anyone think differently, but rather to suggest alternative perspectives that might be helpful for people coming into contact with this notion of 'New Age' for the first time and who are formulating their own ideas and responses.

When someone says 'New Age' what do you think of? In the West the words are often associated in people's minds with complementary medicine, community living, meditation, alternative lifestyles, crystals, healing and personal growth. It has attracted under its banner a wide diversity of people and ideas, some of which I feel drawn to and some of which, quite honestly, I feel distinctly repelled from.

What we today regard as 'New Age' has in many ways become something of a marketplace. There is a trend in many areas towards intense competitiveness as we are promised all kinds of benefits (couched in terms of spiritual, or material, well-being) if we buy this, go there, do that, eat this, drink that, meditate this way or join a particular group. It is not only the so-called 'New Age' groups that seem to offer instant redemption and inner peace. We see it as well in sects linked to some organised religions. These we must distinguish from the true and timeless teachings and wisdom of the world's great religions, truths that too often can get lost behind the misinterpretation and at times deliberate manipulation by people to further their own interests and agenda.

We live in an instant world where we are being subtly conditioned into expecting immediate gratification of our desires. Credit is rooted in the same mind-set. Have now, pay later. Or we might say, "be had for it now, but pay for it later". Sometimes we have to wait for things and, through the experience of expectation, we gain so much more. When we have what we want now and count the cost later, we lose something precious—the joy and thrill of anticipation. We also tend to appreciate things more. We get a more fulfilling experience from something we have truly worked for and looked forward to. It is also my experience that true, sustainable personal and spiritual growth comes for most people slowly and following sustained effort. There are no short cuts or quick fixes on the spiritual path, however glamorous or charismatic the person promising these things to you.

When we are desperately seeking greater meaning in life—and many today are because the world is in a mess and people are trying to make sense of themselves and what they see happening around them—we can be dazzled by the choices and tempted to believe the promises. It seems addictive at times with people endlessly chasing from one thing to another, chasing that illusive high that others are all too prepared to sell or promise us.

Qualities for the new era

Yet how do we choose what is of genuine spiritual depth? How can we discern whether we are really being offered something of quality, something that will truly enable us to develop as human beings? If we were to choose a school, college or university, would we not wish to see the prospectus and the results gained by past students? Would we also not be interested in the manner or style of teaching and the values of the educational establishment? For me, the values are crucial. Whenever I come across something new, some idea or initiative that is claiming to be the answer to everything, I find myself asking:

- Do I experience the presence of a spirit of goodwill?
- Do I find a sense of responsibility finding expression and being encouraged in others?
- Do I experience a note of integrity?
- Can I find an emphasis on group or community?
- Can I sense a spiritual core (realised or not)?

I actually prefer to think about the future in terms of it being a 'new era', a period in the life of humanity that will be full of ideas and opportunity for individuals to encounter each other in a global context. It will be a time for different ideas, cultures and traditions to come together to celebrate and to agree on certain core values that I believe humanity requires if the future is to be sustainable, just and free from exploitation. I want to consider what some of these qualities might be.

The following five factors I think are central to the future:

Spirituality

Goodwill Responsibility

Integrity Community

Let me briefly share my meanings for these qualities. I will return to them in further chapters, but for now what do they mean?

- Spirituality: a controversial word in many ways. For me it is simple. The more human I become the more spiritual I become. I regard spirituality as my essence, my potential and my fulfilment, and it embraces qualities of heart, mind and will.

- Goodwill: I see this as an energy, as an intention to work for that which is good. It is not merely good intentions which can seem a little passive, rather it is dynamic, active and motivating, transforming relationships with a will to create that which is good.

- Responsibility: I believe this is crucial. Without an ethic of personal responsibility we lose social cohesion and we create separation and division. Human Rights are unsustainable without Human Responsibilities

- Integrity: which for me is about how we conduct our affairs in life and the values that we live by. Where self-seeking, exploitative, don't give a damn about the effects attitudes dominate we have lost integrity.

- Community: for whilst we each have our individuality, we need community, we need a sense of belonging, of being participants in the human drama. With community comes co-operation as well. Our first sense of community should be that of the family. Too many do not experience this any more.

When I witness a tendency towards any of the above, I believe that we are seeing a trend into a new era of human activity and relationship. Now, you may be thinking that these are not new ideas, they have been around for centuries, if not thousands of years. This is true and is a point to bear in mind. They are expressions of latent tendencies that have emerged through individuals and societies in the past. However, what I believe is different now is that we have an opportunity of establishing these qualities at the level of pan-humanity. This is the 'new' aspect of the new era ahead of us. The phrase 'global-village' has often been used though I think this conveys rather a rose-tinted image of what the world is really like. But the concept is still useful to have in mind.

What I do believe needs to be emphasised is that the 'New Age' is not just about healing, crystals astrology or meditation, even though there is a strong association with these things in peoples minds. It is actually much, much more and for me the term 'new era' should be associated with a much broader set of values, qualities and attitudes. For the new era is also concerned with demonstrating concern for others in a practical and very down to earth manner: feeding the hungry, sheltering the homeless, clothing the naked and protecting the environment, and doing it with personal commitment, not simply setting up a direct debit and then forgetting about it (though money is of course important). Charity must come from the heart and engage the person in the process of giving and receiving.

Those who are working for human rights, working to make health and social care available to all, working to broaden and strengthen educational standards that truly educate the whole person and not just the mind and memory, are part of a movement towards a new era of human life. Everyone who is working to lift the human condition, individually or collectively, is part of what I would term as a movement, a movement that is taking us towards a new era of human relationship and understanding. And I would add that it in being a part we become a participant, and that is crucial.

It seems there is a real need for reasoned thought when it comes to the 'New Age' marketplace alluded to above. By this I mean thought that weighs up what is being offered carefully against the values and beliefs that are dear to us. Some will say, trust your intuition, and I would agree, but how many of us have actually developed the intuitive sense? Personal feelings are often associated with intuition, an attitude of "I do what I feel" and in extreme without any thought for how it may impact on others. Intuition is rare, but when it occurs it has a genuinely spiritual depth and is unfailing. Personal hunches and gut reactions may be right at times but they are fallible and are not the promptings of the intuitive sense. So where I cannot rely on intuition I would rather use my own reasoning mind to reflect on what is being offered in the light of the questions and keywords listed earlier in this chapter.

There can be no doubt that the speed of travel and communication has changed the face of the world, and the way we perceive the world and the rest of humanity. We do live in a new, global reality, but we probably haven't really fully experienced the significance of this, and in particular the effect it will have and is having on human consciousness. It is a challenging time, and it is something we have to adjust to. It is perhaps worth acknowledging too that more and more people are engaging in processes that will enable them to more fully express their human (and some would wish to add 'spiritual') potential.

The global service network of international aid agencies, the United Nations and its Specialised Agencies, right through to community based organisations and support networks are also part of this one movement, providing a real cutting edge in the world of everyday experience. Whether in the field of human rights, social cohesion, environmental concern, health-care or education, we will find individuals and groups responding to and expressing the key qualities I have highlighted above.

I would suggest that in terms of human growth, the more truly human I become in the sense of realising my fullest potential, the more spiritual I become as well. It seems to me that there is a crossing point between the human and the spiritual. The qualities highlighted previously are expressive of the emergence of that spiritual element, of people connecting with something deeper within their human natures. There is surely some deeper, more significant quality of heart and mind that is present for us all, however hidden it might have become, that increasing numbers of people are yearning and searching for; those that are shaking themselves free from the materialistic and self-centred values that have come to epitomise the prevailing Western civilisation (and other consumer dominated societies) at this time in history.

We have to face facts, and they are uncomfortable ones. The world is in a mess. People die needlessly by the second. The division between rich and poor is obscene. The exploitation of labour in poorer countries is an offence to humanity. The greed that sets people against each other, the lies of the politicians, and media organisations with their emphasis on gossip and sensationalism and which put sales before genuine news that could nourish human intelligence and understanding is, at times, frankly unbelievable. But it is not unbelievable. It is true, it is happening and it holds a dominant position on the world stage. We need change.

Perhaps more contact between cultures, and more personal, human-to-human understanding of the effects of our actions, will make a difference. We must hope so. Indeed, if we do not believe that there will be change and hope for a better world then we may as well carrying on eating, drinking and making merry for tomorrow we will surely die, by the billion, whether as a result of sea level rises, wars over fresh water, wars over gas, further wars over oil, wars over the huge migrations of people that will result from climate change and trade systems that exploit the poor for the material benefit of the rich.

Treading a spiritual path

I believe that many today are experiencing an urge to find greater meaning and for many this takes the form of seeking a spiritual path and/or a field of service. The two go together. Teachings suggest that when we are looking for a spiritual path we will find what we seek when the time is right. What is actually meant by 'the time is right' is when *we* are right, when *we* are ready to take the next step to becoming a more human and, as I mentioned earlier, a more spiritual person.

This next step can take many forms. It may be a meeting with someone which, at the time, can seem like 'chance' but in hindsight rarely is. It could be a book that you come across, I have even heard of them falling off shelves to attract your attention! It might be a mystical experience at a religious gathering. It could be linked to the natural world, a sense of oneness or connectedness with nature and creation. Or it could be some other experience that releases a fresh light on your life and the path ahead of you.

What is important on these occasions is to measure the experience against the highest to which we aspire. Does it resonate to that, or not? If so, test it out. Go with this new idea for a while, live it, and check it out rather as a scientist will test a hypothesis. If it is still OK, then embrace it and truly build it into your life. But always question. Never take the opinion of another as authority. Learn to trust your own sense of what is right, and wrong, for you. Whilst truth may be imparted to us, in a sense it can only become our truth when we have found it for ourselves and built into our lives. We will not live it if we do not own it. We have to learn to trust our own judgements, trust our own evaluation of experiences, or ideas that are presented to us. Read or listen to the words that come to you and judge them on their own merit, not by claims made by the person who produces them. And in an era of 'channelled' material, the same holds even more true.

Look to those who seek or claim to embody the ideas that have become attractive to you. Do they express the values and principles that you believe are reflective of the spiritual life? Are they expressive of the highest that you aspire to?

One of the first effects of committing ourselves to a spiritual path, to seeking a way forward that allows your deeper nature to find expression, is that we often encounter great difficulty. It can seem as though everything is thrown at us, aspects of our nature rise up to block us. Treading a spiritual path is not all sweetness and light, it is extremely hard work requiring perseverance and a steadfast commitment. It requires us to draw on depths of inner strength, probably beyond anything that has happened to us before. Reward lies in success, not personal success measured in property, money or even happiness, but in the joy of knowing that you are living to some purpose, and that the living of it is good.

I personally believe that within each of us is a Soul, a spiritual Self that transcends the everyday senses, and certainly the attitudes of selfishness and separateness that are all too clearly present in our world today. For me, the new era is about opening out our consciousness and finding ways in which we can express our Soul nature. We might think of it as building a bridge, some use the image of a 'rainbow bridge', linking our everyday consciousness to this source of human and spiritual potential. But it is not a simple bridge to build. We have to work on ourselves to build in the sensitivity that will allow us to respond to our deeper, spiritual core. The attitudes of heart and mind that find expression through selfishness, unbridled consumerism and an emphasis on satisfying our own personal desires block us from this inner sensitivity. I see the new era as being about people looking for and finding new ways in which to express their essential human compassion and concern for the welfare of others. The new era involves a striving to establish practical and intelligent methods for expressing our humanity. I would like to emphasise *practicality*. It is vital, I think, that the will-to-serve that is present for so many people finds expression in a form that is truly practical and can make a definite and measurable impact for good. Service, giving of ourselves for the benefit of others, or of the environment, must become a dominant feature of the new era.

Finally, how might we best encourage others to participate in the values for building a new era? Very simply—by example. What will always touch people the most is how we are, how we express ourselves and the values that we are seen to live by.

May your example bring inspiration to others to seek out their own inner, spiritual Self, and their own field of service. In essence, the two are one, for the path to the spirit is through service, and service is the natural expression of the life of the spiritual nature. My belief is that the new era will offer spirituality, responsibility, goodwill, integrity, community and I would add service, as visible pieces of the jigsaw of life.

3

The Wish-Fulfilling Tree

In the previous chapter I touched on the importance of human values and the need to aspire to something greater than personal interests. I would also wish to add compassion and caring as two more qualities which I believe will need to become a much more significant feature within the life of humanity in the coming era. I now wish to tell a story to help illustrate the power of compassion. It is a metaphorical story, and yet it has that human edge to it that makes it somehow real. I hope you like it.

It was a few years ago now that I heard a series of three radio talks that were broadcast by the BBC entitled, 'A Modern Mahabharata'. As I recall they were given by a Professor Purushatma Lal, from Calcutta University. In this chapter I want to share with you the contents of one of these talks because it has been an inspiration to me so maybe it will be to you as well. In some places I am using my own words, in others I use Professor Lal's. I tried distinguishing between them with quotes but it did not work. So I am simply telling the story. I hope he does not mind. Somehow I do not think he will, but I want to express my gratitude to him for the story.

The story is set in India, but it is really set throughout the world. It begins with the return of a benevolent uncle to his village. In the village the children are playing with the kind of things that children play with in Indian villages: sticks and stones and twigs and pieces of string and rag-bag dolls. The uncle is puzzled and he looks at the children with surprise. He asks them, "What are you doing? Why are you playing with these? Don't you know that there's a tree out there, the Wish-Fulfilling Tree, which gives you everything and anything that you wish for?".

Now the children are modern kids and they smile to themselves at these ridiculous words. They know that the world is not like that; it is not a place that gives you whatever you want. They know that even if you struggle hard, there are always those who struggle harder and get the 'goodies' first. And then there are those who have 'connections', who always seem to get more. You cannot compete with them. So they smile at the uncle and continue with their play. But as soon as the uncle has gone they rush off to the tree and begin wishing.

There are two things that children wish for: sweets and toys. They wish for sweets and they get stomach ache. They wish for toys and they get boredom. More sweets, more stomach ache. Bigger and better toys, bigger and better boredom. The children are now becoming confused. Something is clearly wrong. "We make our wishes, and we get what we want, sweets and toys, but what are these unpleasant extras that seem to tag along with them?" Yet they carry on wishing, "at least we are getting our sweets and toys", they say. "We'll put up with the discomfort."

What the children do not yet know is that the Wish-Fulfilling Tree is the hugely generous but totally unfeeling cosmos. It will give you exactly what you desire, that is its job, but with it you will also get its built-in opposite, for this is its job too. Desire always attracts both sides. Nothing in the world has come or will come singly when it is attracted by personal desire. You get the fruits of the tree, and the tree gets you. And so it continues so long as we continue to live by our personal desires, seeking to gratify them without concern beyond ourselves.

Adulthood

Now the children grow up and, of course, their desires change. We call this new phase of desirous living 'adulthood'. The tree has four fruits in particular which hang temptingly from its branches. These four fruits are sex, fame, money and power. How they beckon to us, promising us all that they will make us feel good.

The young adults pluck the four fruits and bite hard into them. They seem tasty enough and so they bite again, and again, and again. Yet in time they begin to find a bitter taste in their mouths, the bitter taste of disappointment. They are somehow not as happy as they thought they were going to be. But they still go on biting and wishing, and eating, and suffering. Well, what else is there to do? What is there to life beyond feeding our desires?

The young adults, who in some ways are still little children, become old. They can be seen sitting beneath the tree in three groups. They are still wishing. The first group whispers, "the world is a farce". Fools, they have learned nothing. The second group murmurs, "the world is not a farce, the mistake is ours, we made the wrong wishes. We'll wish again and this time wish the right wish". Bigger fools. They have learned less than nothing. Meanwhile the third group, the biggest fools of all, say among themselves, "if this is the way the world is then what's the point of living? We wish to die." Gladly and swiftly the obliging tree says "take it" and grants them their death wish. They die, their desire is met, but what is this something other that tags along? Yes, they also receive the built in opposite of death, which is life. They are reborn once more and under the same tree, because the tree is the cosmos and there is no other place to get born into. And so it all begins again.

The parable could end here. It describes how the desire-life binds us to rebirth and to the world of experience and suffering, to the opposites of pleasure and pain. But if it really ended here then it would not show us how we can release ourselves from the clutches of this awesome tree, from being constantly locked into an infinite desire-driven loop of death and rebirth.

The lame boy

The story continues. There was a lame boy crippled by a wasting disease, who had also tried to run to the tree with his friends, but who had been pushed aside in the rush by his eager and aggressive companions. In the commotion the lad had stumbled and fallen, managing to crawl back to his hut where he sat and gazed at the spectacle before him. He was mesmerised and unnerved by what he saw—his friends wishing for sweets and getting stomach ache, desiring more and more toys and getting more and more bored. Time moved as he sat watching. The children grew into young adults and he saw them grasping at the four fruits: sex, fame, power and money, and biting deeply into them; having and being had, suffering and not knowing why they were suffering, making new wishes and finding themselves still experiencing a bitter taste of disappointment.

He saw the group of old men and women, and heard their comments. He continued to watch, mesmerised by the divine comedy, the tragic comedy, being enacted before his eyes. He did not feel superior or angry, he just watched in wonder at what he was seeing. And in the midst of his wonder a gush of compassion arose in his heart for his suffering friends, and in that moment of compassion the lame boy forgot all thoughts of wanting to wish. Oh, he had wanted to wish, but he forgot to wish, he forgot about himself, and in that moment the tree could not touch him. He stood outside of its imprisoning orbit of influence.

Professor Lal makes the comment in the story that the lame boy "had not done the good act … He had not done the bad act … He had done the pure act, the act of gratuitous compassion which has no reward and no punishment. The pure act lies outside of the world's give and take system. The pure act is its own reward and so it is the only way to be free of the all-entrapping wish-fulfilling tree".

Mahatma Ghandi, when asked to explain the pure act is reported to have said, "don't ask me, ask any mother why she puts the baby on the dry side of the bed and puts herself on the wet side, joyfully".

Power of compassion

So the story ends. It conveys the power of compassion whose roots lie in a willingness to share in the sorrows and sufferings of others and to stand free of the temptation of living a desire-driven life. Though we may free ourselves from our own sorrows we should perhaps beware of losing sensitivity to the pain of others. Perhaps we might be more ready to open our hearts to their anguish, to our brothers and sisters around the world who, as we read this, are suffering the worst kinds of atrocity. Perhaps in our genuine moment of compassion we can find that pure act within ourselves through which we can play a part in making a difference in our world.

Can we care enough to accept the hurt that comes from releasing true compassion into our world? It seems to me that compassion is a crucial piece in the final picture of the jigsaw of life. It may be worthwhile simply reflecting on what the final picture might look like if compassion was not part of it. Whilst it is important to dwell on what it will look like when qualities are present; it is also sobering and thought-provoking to reflect on what it might be like if they were absent. Can our world really become what we would hope for, especially for our children and our children's children, without compassion?

Of course there is embedded within this story a belief as well that there is a rebirth. Not everyone is at ease with this view. I hope that for those who read it who do not have this belief that they will also see that story holds for a single life too. And perhaps where there is an accompanying belief in a Heaven or a Paradise, that perhaps the desires of this world may in some way chain us to a less fulfilling experience in the next world, until such time as we reject our own self-importance and embrace heartfelt compassion for others.

Time

Now. What is. Created not by us, it simply is.
Begotten in a suddenness, a moment startled into being by an unknown light.
A knowing look from another world.
Our past is no more, a lost form, yet the memories linger, perhaps to torment us.
But we must move on now—a little closer.
Experiencing a depth of ease born out of a forgotten familiarity
Forms in time, but the time is past.

And ahead?
Onward, upward, no path behind.
Another step to be taken into ourselves
Drawing us still closer until we are
Lost.
Blinded by the limits of our understanding.

And found.

Hand in hand we cross our bridge of time
Linking that which was, and is, and will be.
What will be already is. Of this we can be sure.
The die is cast.

Subtle patterns impress upon an outer world;
Seeds already sown, awaiting the harvest that must surely follow.
Singular fragments united out of time,
Waiting for the moment to bring forth that which Is
Where it is not.

Love.
Only in time can life's inevitability take its final form.
Diverse natures uniting in a common cause.

Rooted in the timeless yet born into the time-bound;
Subtle strands of consciousness woven into the eternal fabric of life.

An inner pattern, veiled by time, is seen in part,
But we must see it Whole
That we may know, in time, what Is.

4

Right Human Relations

As we journey through life we make contact with so many different influences that touch us and, in some way, shape us. It could be a person, or a book, or it may be a film or a play. It could be sacred architecture or places in nature, places where we find ourselves more able to engage with a depth within our being.

In the first part of the last century a spiritual philosopher called Alice Bailey wrote a series of books which, for me, made a very strong impression. In her writings she explores a huge range of topics which include: models of personality, the idea of the Soul and of cosmic evolution, the concept that all is essentially energy, the meaning of discipleship and the future of education. Yet throughout her writings, however complex the world-view that she presents may be, there are some fundamental simplicities that we can all relate to in our own daily life. One of these is the idea of 'right human relations'.

It is my belief that these three words, *right human relations*, are in a sense all we need, if we could only live them out in the routine of our daily lives. They are familiar words, none of them is particularly complex. Yet they seem to present us with such an awesome challenge. What do they really mean? How can we be touched by them in such a way that we seek to let them become alive in our hearts, minds and actions?

We can usefully consider their meaning and application from four perspectives because I think we need to create:

- right human relations within ourselves
- right human relations with each other
- right human relations with other kingdoms of nature and the world in which we live
- right human relations with the Divine

We need to create right human relations within ourselves. Whilst we remain divided within our own natures, being pulled by our emotions in one direction whilst our thoughts pull us in another, we maintain inner conflict which then affects how we are externally. Inner turmoil produces outer turmoil. We cannot find inner peace without first of all establishing some measure of right relationship within our own natures. We might regard this as another way of thinking about having an internal, psychological jigsaw that we are piecing together. (see later chapter entitled *Jigsaw of Self*).

We need to cultivate right human relations with each other. This seems fairly self-evident. Again, unless we treat each other with dignity and respect, and create harmonious and co-operative relationships, then we are going to make our collective life one of constant conflict and struggle. Does life have to be this way? It seems to be the prevailing pattern but I would suggest it is not inevitable that it should be the case that the patterns of the past and the present have to live on into the future. It seems to me that human beings are agents of change with enormous creative potential.

We need to cultivate right human relations with other kingdoms in nature and the world in which we live. We do not live alone or in isolation on Earth. We are part of a vast eco-system that is full of interrelationships, some are subtle, some more obvious and clear. Our choices, our lifestyles impact on the environment and on other forms of life. We have caused a great deal of damage and instability with our short-sighted, exploitative attitude towards so-called development. In the past this may have been out of ignorance, but the era of ignorance is now past. We know that we can adversely affect the environment and threaten not only the well-being, but the very existence of other forms of life in the animal and vegetable kingdoms. We need to transform our relationship with the environment and the lives that we share it with.

We need to cultivate right relationship with the Divine. We will not all feel the need for this, not everyone believes in a Divine Presence. So I do not want my reference to this idea to put you off. My view is that there is a Divine Presence and somehow we have lost something of the inspirational connection with It. Our immersion in material values has cut us off from something deeper that may exist within us (immanent), may exist outside of us (transcendent), but which for me is probably both. Rediscovering some kind of deeper, inner spiritual connection feels important and I think is part of the great challenge we all face, individually and collectively.

Many people are seeking to rediscover spiritual connection, either through organised religion or other approaches. The rise of Buddhism in the West in recent decades can be seen as an example of this, along with the rise today in interest in true Islam. I say 'true' as we must distinguish the true Qur'anic teachings from certain cultural influences that have shaped Islam into their own cultural image and values and which can have the effect of distorting the true teachings. This is, of course, true for some other religions as well. We see the distortion of religious teachings most clearly in those nations which call themselves Christian yet whose prevailing value systems and attitudes on the world stage are clearly not reflective of the teachings of the Christ.

People will take other paths to rediscover a sense of deeper connection. All of the areas highlighted above provide pathways to this. A fuller sense of ourselves through greater inner connectedness, fuller relationship with others and a greater sense of relationship with the Earth and other living things, can all engender spiritual experience.

I would like to turn now to the three words 'right', 'human', and 'relations' and explore them a little further.

Right. What is rightness? How do we know what 'right' is? What is right has to be real and it has to be true. 'Right' signifies something that is appropriate and this has to be set against the idea that there is a purpose to creation. If there is a cosmic jigsaw then there is a final picture to be achieved. What is right is that which moves us towards clarifying that picture. I believe that what is right is to do with living to the highest within us. By highest I wish to encompass all that is good, that is the potential of human beings to function with sensitivity and creativity.

It is acting in a way that we know to be the right thing to do. We just know it. It is a response to the world around us, the events of the day, or the life experiences of others that reflects our higher and deeper potential. It will not be an action characterised or motivated by a sense of separateness, exclusiveness or self-centredness.

So how can we define or describe this sense of rightness? When I look at the colours in nature they seem instinctively 'right'. The mixtures and contrasts work. Sometimes when we hear a piece of music, again we know it is right. The harmony is spot on. It could also be experienced by the sportsman or woman peaking as they enter 'the zone' where they just know they are doing it right. Maybe we are challenged to cultivate, or regain, our sensitivity to rightness. I sense that this is so. And that there is an element of connectedness, of relationship where this rightness emerges into our awareness.

Human. Do we really know what it is to be human? Can we really grasp the full extent that our humanity encompasses? Are we merely biological organisms, our extent no further than our skins, than what we can see? Maybe. Or maybe not. In truth, we do not know. I personally believe (and it has to be a hypothesis) that we are more than this, that maybe there are aspects to our nature beyond the physical body, that we have underlying energy fields that are like bodies of emotional and mental substance. Maybe there are deeper levels too, possibly reaching into what many might refer to as the Divine. Yet we do not need to think only in these terms. What are the limits to our humanity? Do we dare to be truly human in our world, expressing qualities such as compassion without conditions? It seems to me that both our hearts and minds are crucial aspects of what it is to be human. Can they work together? The heart is often seen as being linked to relationship; the mind provides us with the means of creativity. And there is also the quality of will, which gives us direction.

Sometimes we hear 'being human' as an excuse for some perceived weakness or failing. Yes, we are 'only human', and with that comes a range of challenges. Yet it also comes with opportunities and possibilities. We need to be inspired by those who take their sense of being human to the limit, who push the boundaries of human heart, mind and body to show us just what 'being human' is capable of.

Relations. What are relations or a relationship? What energy does it involve? Magnetic and dynamic interaction for a start. There are so many forms of relationship (does relationship have form, or does it create forms?): between individuals, within families, and between friends and work colleagues. These are only some of the ones we are conscious of, however, if at some deeper level we are connected, if in some mysterious way thoughts and feelings do impact on others (energy follows thought), then we are in constant relationship with the whole of humanity—perhaps in some mysterious way with the whole cosmos. Could reality be a continuum of relationship? If so, then we live in an awesome reality.

We are bound by relationship, we cannot free ourselves from it. What do we do with it? How do we act out of the idea that we are all interacting with each other subtly all the time and that, in essence, all that we are and that we create is rooted in relationship?

Can we go so far as to say that relationship is the primary reality? All forms are constituted out of a dynamic system of relationships. Groups, organisations, societies are formed out of relationships. Our bodies exist in coherent form only because of the intricate system of chemical, electro-magnetic, biological and other relationships that underlie the outer appearance. Everything that exists is subject to the power of relationship.

Can we therefore generate, as human beings, right relationship with the world around us? It seems to me that what is 'right' is related to that which tends towards goodness. I see this as a kind of force or impulse, a kind of energy of goodwill. I think that expressing goodwill is a vital part of being human. In fact, I believe that it is goodwill that can be seen as underpinning right human relations, or might be thought of as the energy of right human relations.

I also firmly believe that right human relations are the simple and awesome challenge of our time. So many great teachers—and here we must include the social, mystical, educational, religious, political and many other disciplines in this—have sought to encourage a clarifying of relationship and the establishment of right human relations: with ourselves, with each other, with the world of nature and with the Divine.

I do not believe that this is a vague and mystical idea. This is not the stuff of a utopian dream. It is surely the scientific basis of human existence and potential. It is surely what we must work towards in our own lives, and encourage in others, if humanity is to have any meaningful and fulfilling future and purpose on our planet. We can build right human relations through any of the many diverse yet interconnected field of human activity: education, politics, religion, psychology, science, creative arts, ecology, health, economics and social welfare. Whatever field of action we choose to work in, if we are at our most creative and expressive of our highest potential as human beings, what we are actually focusing is a *will to establish right human relations.*

It seems to me that right human relations and goodwill form another crucial piece, or pieces, in the cosmic jigsaw. If we open our minds to the idea of right human relations and our hearts to it as an ideal to be yearned and worked for, we will change. We will discover opportunities to establish right human relations in our personal world and in the larger world of human involvement. What then may the world be like? If energy does follow thought, if we are all connected in some fundamental way within the energy matrix of creation, if we are all touched by the patterns and attitudes we each live by, then we may discover that we are a much more powerful force for good in our world than we had ever realised. We might also begin to realise that right human relations is not a utopian dream, something beyond our grasp, but a practical possibility. We might even go further. It may well be that the establishing of right human relations as described in this chapter is the only way forward if humanity is to survive and planet Earth is to flourish.

5

Human Rights—Human Responsibilities

"... the sense of responsibility is the first and the outstanding characteristic of the soul"[1]

On December 10th 1948, the General Assembly of the United Nations adopted and proclaimed the Universal Declaration of Human Rights. This landmark document expressed one of the great achievements in humanity's quest for right human relations. It captured something of the mood of the times as humanity, torn by the pain and suffering of War, sought to establish ideals that would usher in a new era of human relationship. They were visionary times.

It is worth viewing this document from a wider perspective. Its formulation was, in many respects, a direct result of the experiences of the 2nd World War and the period prior to that when individuality was to the fore. The world realised that individual human beings had rights and that they needed to be protected. The World War had taught those with political power that codes of practice were required if human well-being was to be safeguarded. Hence we now have the historic statement:

Article 2 Everyone is entitled to all the rights and freedoms set forth in this Declaration, without distinction of any kind....

The succeeding articles go on to clarify what these rights are, for instance:

Article 3 Everyone has the right to life, liberty and security of the person.

Article 4 No one shall be held in slavery or servitude ...

1. Bailey, AA (1955), *Discipleship in the New Age, Vol II*, p 390. Lucis Press Ltd, London and New York.

Article 5 No one shall be subject to torture …

Article 7 All are equal before the law …

Throughout the Declaration most Articles begin "Everyone has the right …" or "Everyone is entitled …". Most define what we, as individuals, should expect both in terms of our own freedoms and in terms of being treated with decency and respect by others. Yet we need to bear in mind that Human Rights do not exist unless we take responsibility for encouraging them. Human Rights, as an ideal, are incredibly important and yet, on their own, they are not enough. You cannot sustain Human Rights without individuals, organisations, societies and governments owning and acting on a set of human responsibilities. It is through our accepted responsibility towards each other that human rights can flourish in our world.

There are two articles that point us in this direction:

Article 1 All human beings are born free and equal in dignity and rights. They are endowed with reason and conscience and should act towards one another in a spirit of brotherhood.

Article 29 (1) Everyone has duties to the community in which alone the free and full development of their personality is possible.
(2) In the exercise of their rights and freedoms, everyone shall be subject to such limitations as are determined by law solely for the purpose of securing due recognition and respect for the rights and freedoms of others and of meeting the just requirements of morality, public order and the general welfare in a democratic society.
(3) These rights and freedoms may in no case be exercised contrary to the purposes and principles of the United Nations.

These two Articles indicate that there is more to Human Rights than ensuring that the individual is free of tyranny, injustice and discrimination. They point out that Human Rights have to be placed within a social context. We read that human beings "should act towards one another in a spirit of brotherhood" (Article 1) and that "everyone has duties to the community …" (Article 29). For Human Rights to be sustainable and to become a basis for right human relations, we have to recognise the responsibilities that we have towards each other, and towards ourselves. We have to consider how our behaviour and attitude contributes to human relationships. We surely have to begin to take responsibility for the effects of our choices, both individually and collectively.

Human Responsibilities require us to think in terms of our contribution to society and to the ideal of brotherhood or, as it was expressed at the time of the French Revolution, Fraternity. They require us to ask of ourselves whether we act in ways that enable Human Rights to flourish, or do we diminish or undermine them?

Towards A Declaration of Human Responsibilities

The following is an idea for a general code of conduct based on the ideal of a Universal Declaration of Human Responsibilities that I formulated in the mid 1990s. The aim was to encourage thinking with regard to the notion that we need to take responsibility to ensure that we encourage well-being and human potential:

- Every person has a responsibility to behave in ways which actively foster, and do not threaten, the well-being of other human beings
- Every person has a responsibility to behave in ways which actively foster, and do not threaten, the well-being of the planetary eco-system
- Every person has a responsibility to respect diversity and individuality whilst working to encourage the co-operative spirit
- Every person has a responsibility to strive towards realising their fullest potential as a human being
- Every person has a responsibility to encourage others to strive towards realising their fullest potential as a human being[2]

It may seem simple, and perhaps that is both its strength and its weakness. Can we truly argue against our responsibility not to threaten the well-being of others, but to actively encourage it; to not only avoid damaging the well-being of the planetary eco-system but to work towards safeguarding and improving it? We cannot insulate and isolate ourselves any more from the effects of our actions and choices. We are connected to each other in the world, our choices affect other people and the choices of others affect us. And some would argue that at a deeper

2. Further information about this Declaration (including versions in Polish, Portuguese and Spanish) can be found at: http://www.bryant-jefferies.freeserve.co.uk/udhr.htm

energetic or conscious level we are also connected in a more subtle and yet more immediate way.

We are linked, in essence we are essentially unity expressed through multiplicity, yet we have managed to confuse ourselves into thinking that multiplicity is rooted in separateness that is linked in our minds to our appearance as separate human beings. We truly live under the shadow of a lie, the lie of separateness. This has to be seen through and dissipated to enable humanity to move on towards fulfilling its greater potential. Increasing acceptance of human responsibility is an important indicator that this lie is breaking down.

This we are witnessing as more and more people choose to change their behaviour in order to contribute to safe-guarding the environment, to ensure that their money is invested in ethical funds or to take the trouble to check that products they buy have a fair trade standard.

When I originally drafted my own version of a Universal Declaration of Human Responsibilities, the topic was very much on the international agenda. It was an idea that was attracting thought and as an ideal it was emotionally compelling. There were a number of initiatives taking place to attempt to get some form of 'Universal Declaration of Human Responsibilities' established on the world agenda.

Universal Declaration of Human Responsibilities—Proposed by the InterAction Council, Tokyo, 1997[3]

This draft document begins with a Preamble that makes the important point that "the exclusive insistence on rights can result in conflict, division and endless dispute, and the neglect of responsibilities can lead to lawlessness and chaos". It goes on to point out that "the rule of law and the promotion of human rights depends on the readiness of men and women to act justly". In other words, people have to take responsibility for acting in ways that promote human well-being and justice for human rights to exist. Indeed, we might go so far as to say that 'human rights' are unachievable and certainly unsustainable as a reality if they are not linked in some way to meaningful 'human responsibilities'. In the final analysis, the notion of human rights is something that humanity has created. There is no 'divine right' for anything. People can say "I have a right to this", or "that is my right", but this is only so where a society has taken responsibility to ensure that individ-

3. This Declaration can be accessed at: http://www.interfaithdialoguebasics.be/universal%20declaration%20of%20human%20responsibilities.htm

ual rights, or rights as they apply to groups of people, are protected by law. And even then, there will be individuals and groups who choose to break the law, believing their right to behave in a particular way is more important than the rights of the person that their actions impinge on.

This Declaration was drafted under a set of headings, which themselves offer a significant focus that we need to be responsive to as we prepare for the future:

- Fundamental Principles for Humanity
- Non-Violence and Respect for Life
- Justice and Solidarity
- Truthfulness and Tolerance
- Mutual Respect and Partnership

I wish to simply quote from the first four articles in this chapter to provide a flavour of what was being said.

> *Fundamental Principles for Humanity*
>
> Article 1
> Every person, regardless of gender, ethnic origin, social status, political opinion, language, age, nationality, or religion, has a responsibility to treat all people in a humane way.
>
> Article 2
> No person should lend support to any form of inhumane behavior, but all people have a responsibility to strive for the dignity and self-esteem of all others.
>
> Article 3
> No person, no group or organization, no state, no army or police stands above good and evil; all are subject to ethical standards. Everyone has a responsibility to promote good and to avoid evil in all things.
>
> Article 4
> All people, endowed with reason and conscience, must accept a responsibility to each and all, to families and communities, to races, nations, and religions in a spirit of solidarity: What you do not wish to be done to yourself, do not do to others.

These and the other articles within this Declaration enshrined a set of values in the context of human responsibility that deserved fuller recognition. I hope that by drawing attention to them in this book more people will read this Declaration and promote the notion of human responsibility.

United Nations

The *Declaration on the Right and Responsibility of Individuals, Groups and Organs of Society to Promote and Protect Universally Recognized Human Rights and Fundamental Freedoms*[4] was adopted by the United Nations in 1999. Whilst the Declaration does refer to responsibilities it does so in a manner that makes the Declaration still, in many ways, Rights-centred, whilst I would argue that it is a Responsibility-centred Declaration that is required. The way the Declaration is written is to make it clear that whilst individuals have rights, it is the State that has the responsibilities.

In introducing the various articles of the Declaration we read that:

> "the prime responsibility and duty to promote and protect human rights and fundamental freedoms lie with the State"

which is then followed by a recognition of

> "the right and the responsibility of individuals, groups and associations to promote respect for and foster knowledge of human rights and fundamental freedoms at the national and international levels"

To quote from the opening articles:

> *Article 1*
> Everyone has the right, individually and in association with others, to promote and to strive for the protection and realization of human rights and fundamental freedoms at the national and international levels.
>
> *Article 2*
> 1. Each State has a prime responsibility and duty to protect, promote and implement all human rights and fundamental freedoms, *inter alia*, by adopt-

4. The Declaration can be accessed at: http://www.unhchr.ch/huridocda/huridoca.nsf/(Symbol)/A.RES.53.144.En?OpenDocument

ing such steps as may be necessary to create all conditions necessary in the social, economic, political and other fields, as well as the legal guarantees required to ensure that all persons under its jurisdiction, individually and in association with others, are able to enjoy all those rights and freedoms in practice.

2. Each State shall adopt such legislative, administrative and other steps as may be necessary to ensure that the rights and freedoms referred to in the present Declaration are effectively guaranteed.

There is therefore indication that the individual has a right to promote and strive for human rights, but the state has responsibility to implement human rights. There could have been a statement to the effect that "Every citizen has a responsibility and duty to protect, promote and implement all human rights and fundamental freedoms", to emphasise that human rights are the responsibility of us all, both individually and collectively.

Club of Budapest

Another initiative that carries a vision of responsibility for our brothers and sisters on Earth has been formulated by The Club of Budapest in its *Manifesto on the Spirit of Planetary Consciousness*[5]. This *Manifesto* emphasises the need for more responsible attitudes in our world. Concerning this important manifesto, and the need to look beyond human rights, Erwin Laszlo has written that:

"A more responsible attitude is sorely needed. *The Manifesto of Planetary Consciousness*, adopted by The Club of Budapest and signed by the Dalai Lama and two dozen other spiritual and cultural leaders, noted that in the course of the 20th century people in many parts of the world have become conscious of their rights and of the many persistent violations of them. While this development is important, in itself it is not enough; we must also become conscious of the factor without which neither rights nor other values can be effectively safeguarded: our individual and collective responsibilities."[6]

5. Ervin Laszlo, *Network: The Scientific and Medical Network Review*, "Consciousness-Creativity-Responsibility", No 65, December 1997
6. Ibid.

He goes on to quote from the *Manifesto*:

> "In today's world all of us, no matter where we live and what we do, became responsible for our actions as private persons, as citizens of a country, as collaborators in business and the economy, as members of the human community and as individuals endowed with an unique mind and consciousness."

The *Manifesto* focuses on certain key areas, identifying our responsibilities: "as private persons", "as citizens of our country", "as collaborators in business and actors in the economy", "as members of the human community", "as individuals endowed with a unique mind and consciousness". In all instances the emphasis on responsibility is extended beyond the individual to the effects the actions have on others:

> "As private persons we are responsible for seeking our interests in harmony with, and not at the expense of, the interests and well-being of others ...
> As citizens of our country, we are responsible for demanding that our leaders beat swords into ploughshares and relate to other nations peacefully and in a spirit of co-operation ...
> As collaborators in business and actors in the economy, we are responsible for ensuring that corporate objectives do not centre uniquely on profit and growth, but include a concern that products and services respond to human needs and demands without harming people and impairing nature ...
> As members of the human community, it is our responsibility to adopt a culture of non-violence, solidarity, and economic, political and social equality both in our own families and in the family of nations, promoting mutual understanding, empathy and respect among people whether they are like us or different ...
> As individuals endowed with a unique mind and consciousness, our responsibility is to encourage comprehension, empathy and appreciation for the excellence of the human spirit in all its manifestations, and for inspiring awe and wonder for a cosmos that brought forth human consciousness—and holds out the possibility of its continued evolution toward a planetary dimension, marked by growing insight, understanding, love and compassion."[7]

These are inspiring words, painting a clear picture of the potential of the human being to move into other dimensions of being and relating.

In many respects the history of humanity has emphasised the development of the individual, though there have been times when groups of people have come

7. The Club of Budapest 1996) *Manifesto on the Spirit of Planetary Consciousness*

together to achieve a particular ideal. The "self-conscious, individualistic way of life" has dominated, and this is in marked contrast to what has been described as a vision of "a way of brotherhood, of love and of group consciousness"[8] which is surely now required both in the present and for the creation of a sustainable future of tight human relations.

Conclusion

We live at a time of individualism, during a period where individual rights are regarded as paramount, and we see the struggle that this causes as each different faction seeks to assert its rights over others: the loggers believe they have a commercial right, the forest peoples a right to their land and way of living; the politicians believe they have a right to hold on to power, the people a right to choose their leaders democratically.

We need to look further as has been emphasised above. We need to grasp the nettle of human responsibility, and if we do perhaps we will discover that it has no nasty sting after all. We shall discover the truth that a sense of responsibility is indeed the first sign of Soul contact, of connection with our deeper selves. The flickering debate today on the topic of human responsibility needs to be fanned into a flame. It will then fire the human imagination and stir the human heart. It will galvanise human will into action. Evolutionary development has correctly put Rights on the human agenda, we must now build on this, take this valuable recognition with us, and add to it. That addition will surely be a Universal Declaration of Human Responsibilities that will, in so many ways, map out the tone and the emphasis of the coming era and which will create the foundation on which a future world of right human relations will be established.

From a spiritual perspective, the sense of responsibility is the first and the outstanding characteristic of the soul because the soul nature does not know separation and self-centredness as we all too easily experience it. There is something so timely and right about a Universal Declaration of Human Responsibilities that I believe it to be truly irresistible. Of course, this does not mean that it is inevitable and we can sit back and let it take shape; it will only do so because people want it and are prepared to actively demand it and to live by it.

I am left with no doubt that in the jigsaw of life the factor of responsibility needs emphasis. It seems to me that it is one of the key qualities and attitudes that will need to be present for humanity to have a chance of creating a sustainable

8. Bailey, A (1942), A Treatise on the Seven Rays Vol II, Esoteric Psychology II. Lucis Press Ltd, New York and London. p. 744

future that dignifies the worth not only of ourselves and each other as human beings, but also of the Earth with the abundance and diversity of life with which we share our planetary home.

◆ ◆ ◆

As I looked down, I saw a large river meandering slowly along for miles, passing from one country to another without stopping. I also saw huge forests, extending across several borders. And I watched the extent of one ocean touch the shores of separate continents. Two words leaped to mind as I looked down on all this: commonality and interdependence. We are one world.[9]

9. John-David Bartoe, US astronaut. Quoted in *The Home Planet*. KW Kelley (Ed) Addison Wesley Publishing Company. (1998)

6

Seeking Equality in a Diverse World

We live in a world in which we are becoming more aware of the global perspective. We are experiencing more direct interaction with a wider range of people and peoples with different beliefs, habits and customs. There is a greater sense of responsibility at large as people witness at first hand, or through the media, the effects on others of the choices that they make in their personal and collective lives. We are becoming in a sense like a global village although I have to say it feels sometimes more like a global shopping mall with those who can afford it inside and experiencing a compulsion to buy, or at the least to look and hope for what they might one day possess. Outside are those who cannot afford to buy, those who may be being exploited within the processes of production, or those who are simply the victims of the political corruption that also contributes to the ruining of economies. The haves and the have-nots. And there are those both inside and outside who are actively seeking to attract the attention of the compulsive shoppers to the plight of the have-nots.

I have introduced this chapter in this way because for many people the reason that they are outside comes down to difference, to the fact that they belong to a culture or a group that is perceived by those who hold the economic and political power as being outside the mainstream. This happens on both a global, national and local scale. There are inequalities within the social, economic, health, political, educational and so many others areas of human involvement. The many inequalities that are associated with difference are, quite frankly, unacceptable, and yet it seems that separative attitudes that are partly linked to fear but generally to ignorance continue to pervade institutions and communities.

In the final analysis everyone is unique. Yet we also experience similarities with some people that lead us to identify with particular groups. It is easy to start to refer to people as different, but that always implies that the group to which they are different and do not belong to is somehow the acceptable group. All groups have differences from other groups. Some will be large and dominant groups and some small. Some will be predominantly discriminatory in their attitudes and behaviours, some will be predominantly the targets for that discrimination.

I want to say something here about our use of language. I used the word 'unique' and I believe this is an appropriate word to use when we are referring to individuals. We are all unique. Yet within the mix of character traits, beliefs, racial identity and other factors that make us who we are, there are going to be elements that are similar to others. Some of these similarities may be such that we either identify ourselves with a particular group, or groups, or we are identified by others as belonging to a particular group or groups. It may not be something that we strongly identify with, or which is a major aspect of our own experiencing of our sense of self, but it could be something that others recognise and which attracts from them a set of negative reactions.

The most obvious groups are those associated with race and ethnicity. We have to face the fact that people are being treated differently because of the colour of their skin, their ethnicity, their nationality, their culture. And by 'differently' I mean in a manner that renders one group at a disadvantage in comparison to another group. This has to change and many countries are passing laws to outlaw racial discrimination. However, legislation alone cannot and never will be the whole solution. It may enshrine values in law, and it may require individuals and organisations to behave in particular ways, but in the final analysis individual people have to think and feel differently. Laws do not achieve this. Education and understanding can. Multi-culturalism, in which a society is made up of a range of cultural groups and in which each group stays separate from others, will not achieve this. Inter-culturalism, where there is a genuine cross-racial and cross-cultural sense of community, can. This requires respect for difference and an ability by people to acknowledge and appreciate each person's individual uniqueness rather than simply seeing people as a grouping towards which a set of negative reactions are then projected. There is a lot of negative stereotyping that takes place and it puts up an immediate barrier to getting to know individuals.

At the same time that we want to encourage people to all be treated equally and fairly, there is a strong wish from people to want their particular culture to be respected, for their differences to be acknowledged and accepted and allowed to flourish without their race or ethnicity being a cause of their being disadvantaged. At the same time, we also need an awareness that race and ethnicity should not be seen as a reason for a group to be advantaged either.

But we start to run into problems. We do see inequalities in the workplace. There is a 'glass ceiling' that means that people from black and non-white British ethnic groups are disadvantaged, do not achieve the same levels of promotion as their white British colleagues (this chapter is written primarily in the context of the author's experience of living in England). How is this to be addressed? Do we allow positive discrimination? At an individual level this is then giving advantage to a person because of their race or ethnicity. Is this truly a way forward? In the wider social context, perhaps it is. If we simply look at the statistics then we would see a shift and we can then say, yes, our society or organisation is now appropriately ethnically diverse, reflecting the population at all levels of seniority. Yet in achieving this, other individuals—those who are, if you like, the victims of someone else's positive discrimination—will feel aggrieved and unfairly treated because they are now not achieving what they believe is their potential because of their race or ethnicity. It is not only an issue in terms of employment and the workplace.

We could say, well, we have to make sure that everyone has the same education and that this allows everyone to have the same life-chances. This might be a way forward, so long as those who make the decisions as to who receives the best education, who passes exams, who is supported most within schools, is racially fair and equal. Yet within some Western societies (and I am thinking of the UK because that is where I come from) there is no doubt that the in-built and dominant white British culture sets a certain accepted way of functioning and this, by its very nature, can set those who approach life with a different set of customs and ways of doing business at a disadvantage. I again use the word 'different' but perhaps I should say 'another', another set of customs and ways of doing business. Does that change the meaning and emphasis?

We need to acknowledge, too, that the equalities agenda extends beyond race and culture. There are other groups who experience disadvantage: people who have a disability; women; transgender people; particular religious groups; gay, lesbian and bisexual people; the older person; people living in poverty or who are homeless and the unemployed. And of course to separate the groups out like this is not always the most helpful as individuals may identify with more than one of these groups.

The Universal Declaration of Human Rights referred to in an earlier chapter gives us a focus for acknowledging that all people need to be treated equally, irrespective of their differences. If humanity is to achieve some measure of right human relationship such that difference does not lead to inequality, then there has to be a radical shift in attitude, and this has to be underpinned by fundamental shifts in both heart and mind.

We hear a lot about tolerance, but to tolerate something or someone is, in effect, 'putting up with'. It is an attitude borne out of feeling one thing but controlling that feeling so as to behave in a different way. Actually, this is a state of incongruence, a mismatch between inner experiencing and outer expression, and it is not a healthy state of being. We have to move beyond tolerance to genuine acceptance of people. It may also mean that we will need to dis-identify from a particular difference, a particular identity, and begin to embrace a stronger sense of being *a human being*, first and foremost. This is our essential commonality and should, I would argue, form the basis for our primary identity. Our cultural and other identities are then secondary to this. A good example of this is nationality. Perhaps we need to think of ourselves less as being British, American, Indian, Egyptian, Nigerian, Chinese, and instead, whilst not losing that sense of self, ensure that at our heart (or should we say 'in our heart'?) is a sense of our humanity, and of our common identity through that sense or feeling of identity with all people, irrespective of differences in outward appearance, custom or belief.

I think it is reasonable to say that if I can see others as first and foremost human beings, then I am much more likely to relate to them as I would wish them to relate to me. But it is, perhaps, easy for me to say this. After all, I am a member of a group of people that tends to be the one—at least certainly in Western Society—least discriminated against and the most discriminating towards others in the sense of treating people unfairly or unequally because of their difference. I'm white, male, middle-aged, heterosexual, able bodied with no allegiance to any particular religion. Yet, ironically, I am also perhaps a member of the group that is the most challenged by the Equalities and Diversity agenda.

Why do I make this statement? It seems to me that Equalities and Diversity is about attitudes, about how I perceive myself, how I perceive others and how I perceive the relationship between us. I have to ask myself what is it in my nature that means that I may relate to others in ways that can have the effect (directly or indirectly) of them experiencing inequality? We have to start with ourselves. In truth, there is no other place to start. We have to know ourselves, our reactions and our perceptions to people who are different to us. It does not mean that I necessarily have to start 'liking' what another person or group of people may do as part of their culture, but I do have to explore how I might accept the other person even though I may have some fundamental disagreements with them.

Just how racist are we without thinking of ourselves as being so? Imagine you are walking along a path next to a river. You hear shouts, calls for help. You look across and there are two people in the river, both separated from each other. They are in trouble and are clearly at risk of drowning. You are a good swimmer and are confident that you could save one of the two people, but you doubt there would be time to save them both. They are equidistant from where you are standing on the river bank. One is a black person, the other white. What is your immediate sense of who you will save? Stop for a moment and consider your reaction to this imaginary situation.

Of course, it might not be a simple black and white matter. It could be that one person is Chinese and the other is an Indian or a Bangladeshi. What then? Or suppose they both appear to be of the same ethnicity or race, but one is male and one is female. What if you knew that one was gay or lesbian and the other heterosexual? What thoughts are with you now? Would that make a difference? You have to act quickly, you have to make a choice. What factors are shaping your decision as to who to save? One is calling for "Allah", the other to "God", has this made a difference to your decision?

We have to be honest with ourselves, about how we think and feel towards people who, for whatever reason, are different from us. To not do so will mean that discriminatory attitudes will remain and will continue to affect the way people relate to each other.

There is an additional layer of complexity to consider, and I made a partial reference to this earlier when I mentioned laws, though in the context of laws that have been enacted in order to outlaw discrimination. There are also national laws that may be in place which cut across or indeed outlaw particular cultural or religious beliefs and customs. Or they may be laws that, in the way they have been formulated, do not affirm that a person be treated equally irrespective of a particular difference in belief or custom. Some countries have laws that specifically outlaw discrimination, though this may not extend to all of the areas of difference and diversity within which discrimination is experienced.

In the UK there has been a shift in emphasis in recent years with equalities legislation moving towards a greater emphasis on a pro-active stance. In other words, not simply rendering discrimination illegal, but placing a duty on organisations in the public sector in particular to not only outlaw discrimination but also to actively promote good relations between diverse groups of people. This is the position with regard to Race, Disability and Gender and is likely to be extended to Age, Sexual Orientation and Faith/Belief.

However, as I said before, legislation can only achieve so much, it is a change in heart and mind that is required. This means interaction, communication, being open to other perspectives and beliefs, listening and sharing. We do not have to become another's difference, we do not have to agree with someone else's belief or custom, but we can develop the capacity within ourselves to accept the other person's uniqueness and the groupings of similarities amongst those who are different to us. And at the same time there may be customs and behaviours that are unacceptable or that may contravene national laws. It seems to me that the laws of a country should be respected by all who live there, and in saying that I am mindful that there are going to be circumstances where there are bad laws that need changing, or new laws that need to be introduced and enacted in relation to the process of seeking equality in our diverse world.

I would like to offer another perspective on equalities and diversity, and it is linked to a notion I have already made reference to in this book. It is the idea of each person having a Soul, or perhaps more accurately that each person is in essence a Soul, an identity that transcends in some mysterious way the strictly human experience and identity, and therefore we could argue transcends the differences that this chapter has been addressing. This essential, we might say core, aspect of our being is a deeper sense of self that transcends the sense of separation and difference, that in a sense regards and knows human beings as having or of being essentially a spiritual essence.

It returns us to the idea that whilst we acknowledge and respect difference, diversity and uniqueness—and it is interesting to note how each of these words can conjure up contrasting meanings—we can also acknowledge a deeper, spiritual commonality. This emerges from a place within our natures which, when connected with, takes us to a sense of self characterised more by connection than separation and by commonality than difference.

> Of such transpersonal possibilities the most enlightened men and women of all ages have given testimony, expressing them in basically the same way, above the differences and colourings due to individual and cultural conditionings.[1]

The presence of this potential also has to play its part in enabling us to address issues of equality and diversity. It offers another perspective. Yes, we need our differences and our uniqueness, yet perhaps we are more than these as well, and we need to be able to acknowledge this too. Otherwise, there is a real risk that difference will be a force for separation, fear and distrust, when in fact what may be more important is that we seek a sense of unity underlying diversity of expression and identity.

1. Assagioli, R (1974) The Act of Will. Turnstone Press, Wellingborough. pp. 125-6

So, in conclusion and returning to the theme of the jigsaw of life, the final picture I am sure requires good relations between people and peoples, irrespective of as well as inclusive of differences and diversity. Yes, we must celebrate, honour and value each person's uniqueness and the differences that contribute to this, together with the various groups of people who are united by similarities within their individual uniquenesses. And we must also not lose sight of the fact that there exists within us all a fundamental commonality, our common humanity, and perhaps deeper still the Soul.

Freedom's Cost

Shattered forms are strewn across the darkened streets of pain
Where hope was born, had died, was born to live in hearts again.
Tortured bodies buried, so many barb-wired bound;
A legacy of evil that for too long did abound.

Visions of a future time when freedom's bell would ring
Had touched the minds of many, but the struggle it would bring
Would test the Soul of those who stood to fight for what was good.
The final freedom they attained was written in their blood.

Nothing True or Real is ever gained without a cost,
And all our strivings that we make for freedom are not lost.
The Will that drives the spirit on to greater liberty
Lies at the heart of that great life we call Humanity.

7

Unanimity: A Way of InterFaith Co-operation

In this chapter I do not set out to in any way try to suggest any one religion is better, or worse, than any other. This chapter is not about making comparisons or judgements on particular beliefs within religious streams of thought. Perhaps there have been a number of prophets down the ages and seers who have been able to present humanity with a deeper vision of life and its Purpose. Perhaps some of those were indeed inspired by God or via angels or Beings of a similar nature but referred to differently in the various religions traditions. Are we to try to bring these beliefs together, or are we best advised to focus on a particular religious system? Is a kind of religious 'pick-and-mix' really the answer or are there other ways that religions can work together for the benefit of humankind and to serve the Purpose of creation?

One area of human life that has struggled to establish a vision of right human relations is that of religion. Even where sacred texts call for right human relations they can somehow get ignored and forgotten when emotions and fears run high. And, of course, there is the risk that religious texts are deliberately mistranslated or misinterpreted to satisfy the interests of a particular group or faction. It seems that in the world of inter-faith co-operation there are clear polarities. It seems that people are either for it, or against it, and there is little middle ground.

One of the great obstacles to faiths coming together in some way is the belief held fervently by many that their religion is the one and only true religion; that spiritual truths as presented in their sacred texts are the only legitimate presentation; that a specific form of worship or adoration is right and therefore, by definition, all other forms are wrong.

Do we really believe that a God whose nature is said to be Love is really going to be worried by differences in ceremony? If there is a God, would not this Being be more concerned with what exists in our hearts and minds, the attitudes and the intent that we bring into whatever act of worship or adoration that we choose to make?

It seems to me that the religious life that stems from a mind-set that is preoccupied with form and structure is always going to struggle where co-operation and understanding are called for. Of course, such an approach is not only peculiar to religion and acts of faith, it can also been seen attached to politics, economics, science and education, for instance. It is the tendency which, in extreme, breeds fanaticism; the "I am right, therefore you are wrong" attitude; the "you must believe what I believe, if you know what's good for you" position. When such attitudes prevail and are coupled with the religious life and spiritual beliefs there is real danger that you become a short step away from inquisition and persecution. I regard this as a tendency towards 'fanatical separation'. It is unhelpful, to say the least.

Is this an extreme view? Perhaps, however we have seen and continue to witness extremes in the world of religion. It is often not the religious texts that cause this as much as their interpretation and application. We must each draw our own conclusions. What I believe is important is that when we choose a religion or a spiritual perspective, before we begin to step along the path that is being offered, we take a really close look at the signpost. Look at those who have travelled along the route that we are considering taking and ask if they reflect the values we seek to live by. Consider whether the truths that they express are those that are 'right' to us. Be sure that our choice is informed. We have minds, and it is important that we use them rather than allow ourselves to be seduced into a blind devotion. Surely we need to emphasise similarities and complementarity and rise above the focus on difference and what comes down to 'holier that thou' attitudes, and fear of 'the other'. Do our religious texts encourage this?

Whenever we set out on a journey it is good to have some idea of where it is going to take us. We may not always be able to know every detail, but it is helpful to get as much of an idea as we can. We would normally consult a map in order to choose the best route for us. Is not the same applicable in life when we seek a path to spiritual truth or to fuller contact with God if that is what we believe in?

One factor that has deeply touched me is the fear that the notion of inter-faith activity seems to encourage among some sections of religious life. It is seen as a threat in some way with people seemingly labouring under the idea that their beliefs will be undermined or swamped by something else that is considered foreign, alien, and not to be trusted or embraced at any cost for fear of being tainted in some way. I find this astonishing. I also find it deeply distressing where people cling to sacred texts as being the only true vision, yet act in ways that are totally the opposite to what those texts seek to encourage. It leaves me wondering if they are serious about their religion, or whether they simply need a belief that they can manipulate to be the excuse for their actions—it is a great ways to avoid taking responsibility to be able to justify your actions because "my religion tells me to behave this way". My sense is that for some people the idea of having a belief or of having a faith is more important than the truth or what that faith actually asks of them. The religious or spiritual life, if it is to have any meaning or relevance, has to be lived. It is about how the person conducts their life, the values that they live by, that matters. Otherwise it is only words in a book. And it has to reflect the truth of that religion, not some distorted version or interpretation of it.

What is belief?

What is the nature of a belief? First of all, by the very nature of it being a belief it is not knowledge. A belief is what we think or feel may be the reality with, I am sure, a certain amount of hope mixed in too. Thought, feeling and hope do not, however, make certainty, yet we see people desperately trying to make a certainty out of a belief. It becomes very difficult for them to accept their perspective as a working hypothesis. A kind of fixation can occur and anything that threatens the belief that they have made into knowledge has to be condemned. As sand, cement and water may produce concrete, sometimes thought, hope and the waters of desire produce a psychological concrete that obstructs our vision and reduces our capacity for genuine spiritual and human sensitivity.

I believe that we really do have to watch out for this tendency. It happens to all of us to a greater or lesser degree. We want to believe things to be a certain way, we cling to the belief making a certainty of it so that it gives our life meaning and purpose, and because we have invested so much feeling, so much of ourselves and our hope in it, we are forced to defend it even when our position is quite unreasonable.

We have to beware of this. It cannot be in line with any religious life that calls for mutual understanding and right human relations. It cannot be a sustainable view where greater co-operation is sought. For co-operation to flourish there has to be flexibility and a fundamental acceptance of diversity. This does not mean that we have to adopt others' beliefs or systems, but we can surely be prepared to be open-minded to their beliefs. In the final analysis, we do not have to choose them for ourselves, and so long as their beliefs do not harm others, then accept them as being a helpful set of guiding principles.

I have a drinks coaster in my home with the words: "minds are like parachutes—they only function when open!". I do not know who wrote it, but it caught my eye a few years back. How can our thought-life develop and our beliefs be informed, if we are not open to inspiration from other sources? If the spiritual path is like a path up a mountain, and Divine Truth is to be found at the summit, then we must accept that there is more than one path and, perhaps, each path sees a different face of the One Truth. Let us be open for the religion that we might be quick to condemn may be conveying a face of truth that we are not seeing.

Unity

The inter-faith movement does emphasise unity. It is sometimes said, however, that interfaith co-operation will produce a unity of religions in which each will lose its distinctive identity. This, I think, is something that threatens people. So let us look a little more at 'unity'. To begin with, I think that what many people fear is not so much unity as 'uniformity'. A unity of form would be a good way of describing this. Why do I find myself thinking of concrete again? Is this what is really to be sought after, a uniformity, a sameness with no space for independent expression of the religious life? Are we considering a kind of totalitarian religious belief across the world? I rather doubt that this is what the inter-faith movement is calling for. And I know it is certainly not what I would want to see. One religious form that is the only religious form for all seems desperately narrow-minded. And yet we live in a single world where surely single spiritual principles apply. If there is one God then all true religions, if they are indeed expressive or reflective of God's Will, or informing us of what that one God requires of us, should be saying similar things, though not necessarily the same thing or in the same way. The image of the spiritual path as being a pathway up a mountain as referred to earlier comes back to mind. Let us not take the uniformity path towards unity. We have read this signpost. We have seen where it is pointing. A strictly imposed uniformity of religious life is not a pleasant prospect.

Earlier I mentioned the notion of 'fanatical separation'. It seems we have two extremes here within the religious life: pluralistic separation and a singular uniformity, neither of which have a great appeal to me. Can we bring them together in some way? They seem to me to appear as a pair of opposites, standing at either end of a scale, two ends of a seesaw, perhaps? Could we seek a balance between them? Or would we have a position of compromise, a kind of trade-off to preserve the balance, whilst tensions remain and people feel they are in a place that is not where they really want to be. So often compromise can mean nobody ends up with what they truly want, and in terms of religion this could be translated into nobody ending up with what they truly believe!

However, I do not think that compromise in the religious life as simply a way of balancing extremes is a recipe for sustainable, inter-faith activity and co-operation. In fact, I think it would be a recipe for disaster as it would eventually break down. We need something more inclusive.

Seeking unanimity

I would like to suggest that we view these two extremes of separation and uniformity as opposite points on the base of a triangle.

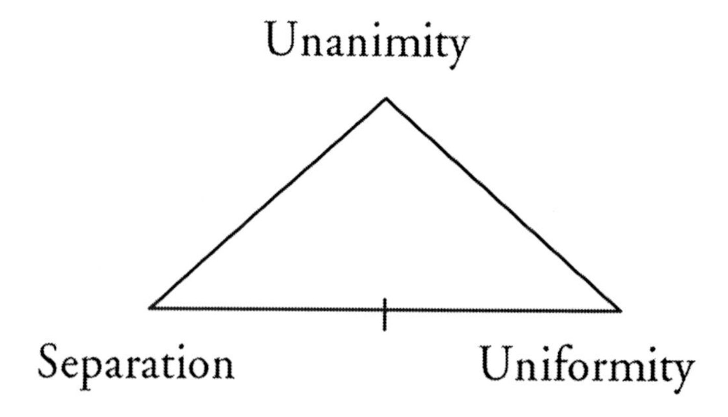

It might be helpful to reflect for a moment or two on how, in this symbolic representation, they might be resolved into an apex. The point halfway along the bottom line is the point of compromise, so what would that top point represent? As the diagram above indicates, I would suggest that the apex is the point of unanimity within the triangle.

Let us consider this word 'unanimity', a word that is not often used and yet which, I believe, presents a more wholesome direction for inter-faith co-operation to take. What does it mean? Unanimity means 'unity of Soul'. It means a unity of intent. It represents a striving towards a similar goal, although we may be taking very different routes to that goal. How might we define such a goal within the religious life? Perhaps a greater spiritual sensitivity and willingness to co-operate with Divine intention and inspiration might be a place to start. There is no doubt scope for great discussion and debate here. With unanimity as our emphasis the focus of unity is seen to shift away from form towards the life of the spirit, towards inner truth, whatever phrase you may wish to use. It is the source of religious inspiration that provides the common focus and not the form through which we give it expression.

Immediately we are struck by a crucial difference—there is space for **diversity of expression**. We no longer have to follow a uniform order of worship or religious practice. We can each reflect our sense of the sublime and our religious beliefs in our own way, as required by a particular teaching or religion. We are free to tread our own paths up the mountain, symbolically speaking. Yet there is not total separation either for there is a recognition that we have a unity of purpose, a unity of Soul, which is to climb the mountain and, of course, that *we are all on the same mountain!*

Some will seek the shortest route, some will wish to be roped together. Others will wish to spend time contemplating the mountain. Someone else will be interested in the view as they climb. There will be those who will be more concerned with assisting others on their climb. Some will wish to mark out a route for others to follow. Still others will wish to climb only at certain times of the day, or of the week. There are bound to be those who will wish to climb alone. There is space for all of these, but we have to reach within us to that sense of unanimity within the religious life for all of this to be possible.

And, of course, there will be those who have found a path and, thinking it is the only path, are desperate to get everyone else to travel with them! Or who are willing to tell people how to find the path, or even how to travel it, for a fee or a donation. I would suggest that we do not necessarily all want to be roped together and be told this is the only way to climb, and we certainly don't want to believe our route is so right that we seek to either brain-wash others to follow us, or worse still try and knock them off the mountain if they insist on taking their own route.

How should we measure the effectiveness of a move from unity through uniformity to unity through unanimity? Why do we need to measure anything at all? It will work if it works. It seems that we are beset by the need to measure and analyse, and it can become a barrier to actually trying things out. The only possible measure of the effectiveness of inter-faith co-operation will be how people work and live together, whether right human relation is fostered, both inter-and intra-faith.

Of course, there will still be those who will warn against this idea, who will wish to cling to either a fear of being swamped by uniformity, or a belief that theirs is the only path up the mountain, join us or else. They will seek to spread fear and alarm, issuing threats of evil influences at work. Let the clanging symbols resound. There is essentially One Truth. It is owned by no-one and everyone. Let us, therefore, unanimously affirm in our hearts and minds the fact of the One Life/One God, whatever names we may wish to use to describe it, that is the Source of, and Which pervades all of, Creation. And what is this One Truth? We are back to the picture that will emerge when the Cosmic Jigsaw is completed as described in an earlier chapter.

It seems to me that unanimity is another important piece in the Jigsaw of Life. Without unity of Soul, without being able to respect, value and celebrate diversity in the religious and spiritual life, what kind of world will we have created for future generations?

8

A Prayer for our Time

Some years back I came across a prayer, or perhaps it is better to call it a mantra or an invocation, that made a deep impression on me. It is called The Great Invocation[1] and I feel that it captures the essence of so much that is pertinent to our times.

<u>The Great Invocation</u>

From the point of Light within the Mind of God
Let Light stream forth into the minds of men
Let light descend on Earth

From the point of Love within the Heart of God
Let Love stream forth into the hearts of men
May Christ return to Earth

From the centre where the Will of God is known
Let purpose guide the little wills of men—
The purpose which the Masters know and serve.

From the centre which we call the race of men
Let the Plan of Love and Light work out
And may it seal the door where evil dwells.

Let Light and Love and Power restore the Plan on Earth

---

1. Available from: Triangles, Suite 54, 3 Whitehall Court, London SW1A 2EF; Triangles, 120 Wall Street, 24th Floor, New York, NY 10005, USA

It is a prayer which, like many sacred texts, has meanings that lie beyond the words. It is said that the words are a translation of inner, mantric sounds. I am in no position to prove or disprove this. I take this on faith and it is enough for me to feel discouraged from considering altering the words as this may unwittingly distort some deeper meaning. However, I also appreciate that not everyone is at ease with the words and I will come back to this later in the chapter. There is an adapted version which I reproduce here for your consideration.

The Great Invocation

From the point of Light within the Mind of God
Let light stream forth into human minds.
Let Light descend on Earth.

From the point of Love within the Heart of God
Let love stream forth into human hearts.
May the Coming One return to Earth.

From the centre where the Will of God is known
Let purpose guide all little human wills—
The purpose which the Masters know and serve.

From the centre which we call the human race
Let the Plan of Love and Light work out
And may it seal the door where evil dwells.

Let Light and Love and Power restore the Plan on Earth

In this chapter I want to share some of my thoughts and feelings on this prayer from more of a psychological angle. In the first part of the chapter I will focus on Light, Love and Will. In the second part I will address some of the concerns that have been raised about the original wording and how it has been interpreted.

Light Love and Will

Many psychological problems stem from an inability to relate wholesomely to circumstances, to others and to ourselves. I do not use the word 'wholesome' in some too-good-to-be-true kind of way, but rather to denote a more holistic way of interacting with others and life in general. Today, inter-and intra-personal relationships are being explored by more people than ever before, and particularly within therapeutic settings. The blocks to wholeness are being revealed as people travel their own inner paths of discovery, seeking higher potential and breaking through self-centred barriers to psychological growth. The idea of the transpersonal realm of consciousness attracts increasing interest and attention. Here we move beyond our own personal sense of self to enter something more profound and universal. To reach this point I believe our human, separative tendencies need to be recognised, a prelude to seeing them as limiting, as aspects of ourselves to be outgrown in order that we might become a fuller functioning human being.

What brings these separative elements to the surface? Striving for greater self-awareness and self-knowledge can do this. A heartfelt yearning and aspiration to reach into the transpersonal realm in a deep and meaningful way can also bring blocks to achieving this sharply to our attention. In essence, we are talking about the invocation of greater light, of releasing a light of understanding on to ourselves. What greater light can there be than that which emanates from the Mind of God?

Not only do we need more light on our path (individually and collectively), but we also need more love in our world. Today, self-love continues to dominate. Love of material form and comfort creates imbalances within the pattern of human living. The cult of materialistic consumerism (is it a cult, or has it now become too mainstreamed in many societies to be considered a mere cult?) encourages this and takes us away from deeper spiritual values and focus. Consuming becomes an addiction. If our expressed love could be adjusted and refocused through a move away from separative and selfish attitudes, from feeding our love of self and love of what *we* want to have and to experience, then we might see more sharing and co-operative human interaction. More and more people relate deeply to this issue. Fair trade, environmental impact, ethical investment are concerns for many people. We see the misery in our world. We see people being damaged by the actions of large organisations out to exploit labour and resources for profit. We know that our purchasing choices support the supply and production chain across the world.

What sensitises us to much of the misery in our world and stimulates an urge to want a more just, caring and equitable society? What makes us want to think about what we buy and the effects of our actions on others? Surely an awakening heart. It brings a sense of human solidarity. We are seeing this and we need more if it. In so many ways humanity requires a change of heart if we are to create a just and fair world. The awakening heart is an invocation of love. Let human hearts awaken and greater love stream forth into them.

It seems to me that we all have to find a way of moving on from pre-occupation with short-sighted, self-centred concerns. Potentially we are 'one-humanity'. Maybe we are in fact, we just have not yet fully realised it in consciousness and assimilated it as an accepted truth to be lived by. Within the framework of modern life we have gained vast amounts of valuable knowledge about ourselves and about our world. Communication systems exist which virtually annihilate time in the transfer of information. It has opened up immense opportunities for sharing knowledge and organising collective action for the common good.

Men and women of vision who care about people and the planet are to be found all over the world and in all walks of life. Yet, even with all of this and with so many inspiring initiatives taking form, what we so often seem to lack is the will to carry vision into action. Many are frustrated and want to act powerfully in this way, their minds awakening to fresh possibilities; their hearts opening to the plight of others and of the planet itself. Yet world problems can seem insurmountable, and the force seeking to maintain the current value systems that dominate the flow of resources in our world can seem so immovable. We can feel too small to make a difference. We need to invoke a greater Will to drive us forwards both individually and collectively. Spiritual will is not something often referred to. For it to find expression in the world it needs human hearts and minds to be responsive to a more inclusive vision and identity than the separateness that currently dominates. I would go so far as to say that all that is working to maintain a separative attitude and the values that are rooted in self-centredness in our world are barriers to the expression of spiritual will. Whilst we cannot expect to understand the nature of the Will of God, we can strive to align our wills with this universal impulse.

Light—Love—Will. These are surely at the core of what humanity needs today, what might be described as the fundamental energies seeking expression through humanity at this time. They are also central to the Great Invocation. I do not believe these are abstractions or "pie in the sky" concepts. Spiritual Light, Love and Will surely lie at the heart of our future. We need to invoke these great aspects of creation, perhaps as never before, to help draw humanity into alignment with Divine intent, and to use the metaphor of this book, to aid us in completing our individual and collective parts of the jigsaw of life.

The Great Invocation's wording

Not everyone is drawn to the Great Invocation. Prayers are quite personal and they need to feel right for us, and they need to capture what we want to express or ask for in ways that we can relate to. I have heard areas of concern expressed about the words that it contains: the gender issue; the idea of sealing evil behind a door; the belief that invocation implies duality and how can this be when all is one; how can Christ return when He is always with us?; the word 'Christ' is too Christian-orientated in a multi-faith world; concern over the idea of restoring the Plan on Earth. All are to be taken seriously and I do not pretend to have definitive answers. All I can do is share how I respond in my own heart and mind to these important issues.

Gender

The masculine focus raises concern, very deep concern, among many people. For me it is an issue of identity. Like many people I have had to tread a path towards finding my own identity. Now, it is possible you may wish to prejudge me because I am a man and whilst I do not and cannot know what it is to find identity as a woman (unless we agree on past lives and that maybe I can encounter the experience of being a woman in a past life and carry some of that awareness into a future life), I do know what it is to find identity as a man, an identity which is acceptable to me and which is not simply a mirroring of media stereotypes or other people's expectations. And I do not like to be labelled as something I am not, or to have the identity into which I have put a great deal being demeaned, as I am sure it is for many others, both male and female.

My own searching has brought me the realisation that there is a level at which the issue of male/female drops away—what I would refer to as the realm of the transpersonal, or of the Soul. This aspect of ourselves, which many regard as being our true selves, transcends the gender issue. In saying this I am in no way belittling gender, it is who and how we are, and justice and respect for a persons worth irrespective of gender is basic and fundamental. I simply want to offer the perspective that there is perhaps something within each of us that goes beyond a gender identity. It is from such a perspective that I seek to use the Great Invocation.

Sealing the door where evil dwells

Sealing the door where evil dwells causes some psychotherapists to throw their hands up in horror. "You can't suppress the shadow!" I have heard said on more than one occasion. I entirely agree with this. If I am to know myself and to be myself I have to know all of me: warts, shadows and all. But as I understand it, the Great Invocation is not talking about evil in the sense of a personal shadow (and I do not believe the shadow that some schools of psychotherapy refer to is 'evil' anyway). Rather, I think this prayer is more concerned with sealing a planetary door to some greater malevolent force operating within the cosmos. I believe that the Great Invocation is of far greater significance than our own personal shadows. If God's Plan for humanity and indeed for the world is allowed to work out into expression then that will seal the door to this kind of evil. Creating right human relations will be an outer and visible sign of that closing for it will mean humanity has triumphed over the sense of separateness and gained (some would say 'regained') an inclusive identity. The door will be closed for there will then be no focus within human hearts and minds for greater evil forces to take root.

From another perspective the Great Invocation actually does have meaning for the process of coming to terms with the shadow aspect. It emphasises the invocation of Light. One effect of increased light is, I would suggest, a greater awareness of all aspects of ourselves, and this includes what some schools of thought refer to as the shadow aspect. As a result of greater self-knowledge and self-understanding we are given the freedom to choose our own unique way of being in the world. Greater light can, of course, be extremely disturbing and unsettling. It may bring us knowledge that is uncomfortable, challenging us in deep and profound ways. We have to make choices. At times, we have to make uncomfortable choices in order to create and assert new patterns of living and being. It is risky, uncertain and often painful. The words of the Great Invocation do not, in my opinion, suggest we create some barrier to facing up to this experience; on the contrary, in my experience it is far more likely to bring our imperfections and limitations to our awareness.

Another angle on this is that if we view the shadow aspect as those aspects of ourselves that we are unaware of, those elements within us that we are currently unable to congruently engage with and express, or which are denied to awareness, then we are forced to consider that the Soul aspect too is part of the shadow side for those who are currently unresponsive to its presence. Hence the importance of thinking of the shadow as being something other than darkness with the often evil meaning attached to it. To open ourselves to more light and love is to open ourselves more fully to the Soul nature which, if it is previously unknown to us, involves an opening up to contents of what is, for that individual, in the shadow.

We have to be careful or else associated meaning that we have with words will confuse us. We have duality within the shadow just as we have duality within the persona, in the sense that we have that which is labelled as 'good' and what is labelled as 'bad'. As far as I can see human/spiritual development is about seeking to evoke from within our natures our highest potential, bringing ourselves as fully as possible into our awareness. If we consider once again the idea of a triangle, we might think of the known and the unknown within us as being represented by the two base points, whilst the apex represents our deeper, spiritual sensitivity, the higher potential that is discovered as we free ourselves from partial identity and become more whole.

We are seeking greater wholeness, greater self-understanding and less fragmentation. We are seeking to know ourselves to the point that we can choose how we wish to be, but from a truly informed perspective. As we respond more fully to our deeper natures, what I am referring to as the Soul nature, and through this become more sensitive to the great potencies of light, love and will, we will be challenged by our own duality. We will come face to face with a stark choice, and yet one that is of the greatest significance in the life of a human being: whether to continue to be dominated by a narrow and separative sense of self, or whether to work towards fostering and strengthening a more inclusive and expansive sense of self.

We are left to reflect on how we each will integrate the shadow aspect. If we take the view that the shadow side of our nature is all that obstructs greater light, love and spiritual purpose finding expression through us, then we might consider the symbolism of standing before a candle in a darkened room. Behind us will be our shadow cast because we obstruct the light. As we move towards the candle the shadow looms larger and larger until ... if we were able to enter into the candle flame the shadow would vanish. It is my belief that the way forward is to keep focused on the source of the light: the Soul nature. Move steadily in that direction and although the shadow may grow, it will cease to be a factor at the point that we enter and become the greater light that is the Soul. I believe the Great Invocation affirms and stimulates this process of moving towards the source of light both for us as individuals, and for humanity as a whole.

Having said that, I want to make clear that this is not the only prayer or affirmation that achieves this. Many religions and spiritual approaches employ prayers, affirmations and invocations that are designed to call on Divine intervention.

Immanence and Transcendence

Let us now turn to the issue of duality, to the idea that the Great Invocation suggests we are somehow separate from God. I have been asked the question: "how can we invoke light and love from the mind and heart of a Deity that is apart from us when in Truth all is One and God is within each of us?" Is the light and love of God truly already in our minds and hearts? If it is then why do we not act as God ourselves, or at least as clear and undistorted expressions of God's intelligent, loving and purposeful Will? Well, maybe some people think they are Gods, but do they act with the wisdom of a God? Rarely, if ever. Surely we do not act as God because our hearts and minds are not yet fully responding to that greater Heart, Mind and Will. At some deep level perhaps all is One as the mystics testify, but until that sense of Oneness is held in consciousness in everyday life and allowed to find expression through the choices and actions of individuals, it will remain separate from us.

I believe we have to invoke Light and Love from that greater Source to stimulate and awaken us. Only when consciousness has expanded sufficiently may we touch that greater Heart and Mind. For me, God, Allah, or the Source of creation, however you wish to think of Deity, is not finding expression in our world. Yes, God may well be immanent within each of us, and at the same time a transcendent Being above and beyond our worldly concerns, but we need to connect with that Source of Life for It to have significance on Earth. We are the links between Heaven and Earth, between the spiritual and the temporal realms, however you wish to define them. If there is a God, and if that God has a Plan for Humanity, it is us who must discover what it is and apply it. We have been given the teachings of the great Prophets down the ages, each offering systems of law for us to live by. Some we have distorted beyond all recognition, some we have cherry-picked, some we have declared too dated to have application today. Perhaps there is a need for an update, for a new set of values to live by that will bring us individually and collectively closer to God, or closer to our deeper, spiritual selves, that build on what we have already been taught? Or perhaps we should focus simply on applying the truths and guidance already given to us.

It seems to me that the need for invocation remains and our choice of emphasis must depend on our own individual experience and preference. May you pray to God in your way, and I will in mine, and perhaps together we might then let God in both His immanent and transcendent state find fuller expression through us and through humanity.

May Christ return to Earth

The idea of the return of the Christ troubles many people too. It can seem too Christian, first of all. And yet many do await the return of a Messiah, of a Great Leader who will lead us from darkness to light. I think that one of the great distortions of religion is the way that Christianity has, in my view, affirmed Christ as their own, and said that He is the only Way to God. Perhaps there is a quality of Being called the Christ that is closer to God. I would differentiate this from Jesus the man who, I would suggest, was touched by the Christ spirit and gave it expression on Earth. Is the Christ then to be likened to an angel, such as Gabriel, conveying the words, the Love and Will of God into the mind and heart of a man for transmission to humanity? Was this how God spoke as well to prophet Abraham, and to prophet Mohamed? Let us consider the Christ as being more than simply Jesus, that what we call 'Christ' is a representation of what we might term the angelic host through which God speaks to humanity.

The Great Invocation implies that the Christ has left us. Some spiritual teachings emphasise the reappearance of the Christ in terms of an emergence into a fuller and conscious recognition of the Glory of God. Yes, the Christ has never left us, He remains "closer than hands and feet". But are we conscious of this? Does the presence of the Christ find expression throughout our daily lives? What I think is important is that we discover the Christ within ourselves and as I see it, what I am referring to here is the Soul, the Christed nature that lies within us all at a transpersonal level of being. It is this that we are destined to draw more fully into expression. It is this aspect of our natures which I wholeheartedly believe must find expression if we are to enter the future with any hope of realising our highest potential. It is this which must return into the midst of humanity, and this time into fuller expression throughout humankind. And it is through the Soul nature that we will find true communion with God.

Let us for a moment then think of Christ as representing something more than a Christian belief. Let us suppose that the Soul nature within each of us is an embodiment of the Christ nature which itself is an expression of the Intelligence, Love and Will of God. Therefore, there will be those who walk the Earth who are more fully conscious of, and connected to, this deep aspect of human nature and potential.

Perhaps 2000 years ago this Christ quality appeared on Earth in a purer form than ever before. But perhaps we distorted the true teachings and created a powerbase called the Church. We know what that Church was capable of in the name of Christianity which was a far cry from the gentle teachings of love that prophet Jesus gave to the world. It has been suggested that Islam was an attempt to swing the teachings back to their essence, the words this time written down and preserved rather than relying on second or third hand accounts of what was said and done. Perhaps there is truth in this, however we are also faced with having to interpret and apply the teachings of prophet Mohamed (as indeed with any other religious teachings) that were written in one historical and cultural context and then translate them into the diverse cultures and societies of the modern world. This is a great challenge, and, as is the case with so many religions, there is the added difficulty of different groupings or sects within each taking different views or interpretations leading to divisiveness and, as we have seen in history and the present, bloodshed.

What is needed today? Could an individual, or group of individuals, exist now to carry fresh spiritual impulse into the world and provide humanity with another opportunity? I recognise that there are many who believe that the Christ will actually reappear in physical form as an individual being in our world. Maybe this will occur too, through someone who is so in touch with the Divine within themselves that they can truly carry Greater Truth into the world. We are already hearing the shouts of "lo, he is here" and lo, he is there". There are those who claim to be in closer touch with Him, claiming spiritual status and promising it to those who will believe and follow them.

I do wonder if He were to reappear on the world scene today whether I could recognise such an embodiment of the Christ Principle if the Christ in me was not sufficiently awakened, at least in some, small measure. I think I probably would not. But then, I am not too concerned with tracking down World Saviours anyway. I will seek the Kingdom of Heaven within me, to strive to enable the Christ impulse (my Soul nature) to find expression a little more in the circle of my own daily life. That seems to me to be more than a lifetime's work! Besides, if the Christ is to reappear in physical form I am sure He would rather I was going about serving human need rather than sitting in adoration, expecting Him to solve the world's problems for us. I wonder if perhaps, if He is destined to walk amongst us, He will only be able to do so when sufficient hearts and minds are awakened to the Christ impulse within us all? Perhaps it is for us to bring the Christ into the world—literally. Or should I say, it is for us to bring God's Love and Purpose more fully into the world for in the final analysis that is what really matters.

The Masters

I personally believe that there are those who have travelled further along the spiritual path than others. I acknowledge this is not everyone's belief. For me, whether I think of them as the prophet's that have gone before us, or the angelic host, the Gods and Goddesses of some traditions, I just sense that those who have given inspiration and direction to humanity do not just move on and leave us. Personally, I do believe that life goes on beyond the death of form, that within us there is a Soul that lived before birth and will continue to exist in some manner when our physical body is dead. I may be wrong, but this is my current view. I will find out when I die, if I die. I cannot prove anything one way or another though I do sense that the presence of people lives on.

But if there is continuity in some form—and it might be continued existence in Paradise or Heaven, or it might mean a return to Earth in another body, but either way it will be through some mysterious process that we do not yet fully understand—then it is reasonable to suppose that the great Teachers and inspirers of the past will remain great Teachers and sources of inspiration in the present and in the future. It seems therefore reasonable to me that maybe they can reach human hearts and minds even when they are not in physical form. If we believe in life after death, whatever we may mean by that phrase, is it not reasonable to believe also that the individuals who have made such great impact on the course of human life may continue to offer guidance and inspiration today?

For me, the reference in the Great Invocation to 'the Masters' is to those who have attained levels of spiritual development such that they are more in touch with the Divine Plan than our less developed human hearts and minds. I see them as bridges in consciousness, offering us the promise of what we ourselves might become. They are sources of wisdom and inspiration in their own right, and can be considered to be links within a great chain of consciousness that stretches the length and breadth of creation. They are therefore likely to work themselves only with those whose minds and hearts can respond to their inspiration, and in particular those who are also in positions of influence to make a positive difference in the world.

But I want to make something clear in order not to mislead. I do not see the Masters as a focus for worship and adoration, any more than I see Christ, Buddha, Mohamed, Krishna or any other prophet as a focus for worship. We worship God, and God reaches our hearts and minds through the Divine Intermediaries, as we might describe them, who can respond to God's Heart and Mind, but who can in a sense step the potency down that we might grasp what is required.

There are many people who claim to be in touch with these Masters of the Wisdom. We must each make up our own minds as to the truth of these claims. For me, I base my conclusion on the quality of any teaching produced, and the responsibility that the individual concerned takes who writes or speaks it. True spiritual teaching has depth and purpose, it is affirming but not in a personal sense, rather in a deeper, spiritual sense. True spiritual teachings will not encourage us towards self-interest or encourage a sense of separateness. They do not encourage us to feel 'chosen' in some 'special way'. Treading the spiritual path is tough and not for the faint hearted. If spiritual teachings do not challenge you, and in particular challenge you to break free of the self-centred patterns of living and thinking, often feeling and desire driven in a strictly personal sense, then it is unlikely to have much spiritual merit. And if it simply demands unswerving and blind obedience to some individual, often involving the whipping up of uncontrolled emotion, that too is unlikely to be true spiritual teaching. Beware, too, of the false prophets who are driven by the profit motive.

Yet I do believe that individuals can be touched by ideas from greater minds without even knowing it, We feel inspired by a thought and we are moved to act on it. Where does it come from? We may not have any sense of having created it ourselves, but it is suddenly there. Do we really know where ideas come from? We experience them as being our own, but are they really? Are our minds brain bound, simply the product of electro-magnetic forces and chemical flow? Are ideas simply a by-product of such interactions? Or can we pick them up in some way because they are alive in the atmosphere around us? We can experience emotional climates, we can sense it 'in the air', so to speak. Why should we not also be able to sense ideas? Do we have ideas, or do ideas have us?

So, to return to the theme of this section, are there great minds at work, both seen and unseen, strengthening the ideas that serve God's Will and Purpose and to which we must respond in order to ensure God's Will and Purpose are fulfilled on Earth? Some will say it is far-fetched and dismiss it out of hand. Maybe they are right. Personally, all I can say is that I don't think they are right to do this and I would prefer to be open-minded to the possibility that thoughts and ideas are being generated all the time by Those who know more than we do. I may not be able to see into the Mind of God but perhaps I can respond or at least glimpse into the minds of those who can and have.

Restoring the Plan on Earth

Finally, restoring the Plan on Earth. This implies that the Plan was established once but has somehow been lost. What do we mean by a Plan anyway? For me it is the picture on the front of the box that contains the Jigsaw of Life. By this I mean that underlying cosmic evolution there is a plan and a purpose. Personally, I don't think evolution is haphazard, however I do sometimes wonder if humanity has in some way 'lost the plot'! Has some plan for humanity been lost? Did God envision how humanity should evolve? Have we lost our way, maybe plunged into deeper materialism, separation and self-centredness than was intended? Are we at risk of trying to jam the pieces of the Jigsaw, and of Life, in any old way without any sense that there is an overall picture to be achieved that is envisioned within the Mind of God? Who knows? I cannot prove or disprove it one way or the other. But I do just wonder if we have become so enmeshed in material values and the endless pursuit of happiness through having more (and in the process causing others who are exploited to have less) that we don't care about the overall picture, just so long as we can satisfy our own personal desires and the values that our consumer society tells us are all important.

Maybe at some time in the past there was a greater connection between the spiritual or angelic realms and the world of everyday human existence. Perhaps at some stage in the past there was a truly golden age of wisdom and right human relations. Perhaps humanity was far more in tune with the Plan at that time and maybe the development of human selfishness and separateness has cut us off from this precious, heavenly state. Perhaps this was a necessary part of the Plan anyway, enabling us to develop and fully identify with a sense of self-consciousness. Could it be that evolution, the divine Plan for humanity, requires us to freely and consciously choose to reject separation as a basis for living, and to freely and consciously choose a way of being that reflects unity and understanding in our diversity of expression and identity? For me, the Plan somehow holds or is the blueprint for righteous living. Is it not reasonable to suppose that original righteousness predates original sin? I think so. God saw that creation was Good in the first instance, so maybe we can work to Let the Plan be restored on Earth, and maybe see that it is Good once again.

I believe the Great Invocation is a prayer for our time. It expresses the hopes and the needs of our world today. We need light on our individual and collective paths; we need more love in the world; we need a greater will to serve a purpose beyond our often selfish aspirations. We need God's Plan restored on Earth. In terms of the jigsaw of life, it represents our need to look towards the Divine, to pray, to seek ways of opening ourselves to higher and deeper inspiration, to open our hearts and minds to the bigger picture. It is concerned with developing the capacity to glimpse something of what is envisioned by God for humanity, and if we cannot see this directly, perhaps we can see it reflected through those who have progressed further upon the spiritual path, who have given us teachings that are, in a sense, the signposts that tell us the direction in which we should travel.

9

The Ant and the Fly

This chapter carries a rather strange title, yet as you read it I think you will understand why. It all began a few years back when I was fortunate enough to spend a week at a summer camp in Poland. An experience there brought home to me how it is the case that when seeking to tread a spiritual path, lessons and learning arrive from a variety of strange and unexpected angles, taking diverse shapes and forms. We always seem to have to be extra vigilant, maintaining a heightened sensitivity to experiences and opportunities that they may provide us with new insight into ourselves and the processes of life.

Anyway, there I was, sitting in the extremely hot sun on a concrete patio listening to a lecture being given in a language that I did not understand. The topic of the lecture for the purpose of this chapter is not important, but I was feeling a bit out of it. I could hear all these sounds that were words in another language but they meant nothing to me. I was probably feeling a bit bemused as well with people all around earnestly taking copious notes. It struck me that this learning business seemed like a lot of hard work, everyone had such serious expressions on their faces.

I guess I must have just allowed myself to drift and to let my attention wander to the surroundings. Suddenly, something caught my eye, something small moving somewhat erratically across the concrete near to where I was sitting. An ant! An ordinary little ant, yet this particular ant was not alone for it was struggling along with an enormous burden—a dead fly. It was clear that it really was struggling, the ant was definitely so much smaller than the fly. For some reason it captivated me, holding my attention and I felt an intense fascination arising from within me. My whole focus, my whole being, somehow found itself focused on the ant as it sought to heave its deceased burden across the sun-baked concrete.

Looking back on it afterwards it struck me that I had in a sense entered into a state of meditation. When we talk of a meditative focus we may think in terms of a seed thought, or some geometrical symbol, or perhaps a mantra. Yet for me the ant and its struggle became a living symbol. What was interesting to me was that I was not fighting to keep my attention on the ant, I was not 'doing' meditation, I was absorbed in a meditative state. It felt as though I was somehow held in a state of wonder. It was almost childlike, as if lost in a world that I did not understand yet which was overwhelmingly impressive. And yes, it was all centred around this lonesome ant staggering around in the hot sun under the dead weight of the fly.

A thought struck me so sharply that it made a profound impact on my mind. I suddenly saw the ant as representing human beings. And the fly? Well, the fly represented knowledge, the kind of knowledge that we seem to have an urge to acquire for its own sake yet which in truth has little real significance in our lives. The ant, symbolising the human being, had a huge amount of knowledge, symbolised by the fly. But it was reeling around. It really did not seem to know what to do with its burden. It seemed to have no purpose or sense of direction. Yet the ant would not let go of the fly. It hung on grimly.

My attention shifted a little and I became aware of other ants, but none came over to my ant to help. This struck me as unusual, ants are usually communal and if one has something for the community they will usually all help. But not in the scene being enacted out before me. They seemed to be shunning this lowly, overburdened ant.

It struck me that even if the ant got back to its nest, it probably would not be able to get the fly into the ant-hole. So why did it bother carting it around all over the concrete in the hot sun? Instinct? Perhaps, yet instinct generally seems to drive action towards some purpose. There seemed little purpose in what I was witnessing.

I began to think about how human beings deal with knowledge. Do we instinctively seek for and cling to knowledge and information, to our own ideas that we think are so important? Maybe.I think that often when we do take hold of knowledge or information we can fanatically cling to it. Could it be that, like the ant, we can also be shunned if we cling to what we know in some intense manner, shunned by those who prefer greater open-or wide-mindedness?

I was left wondering. Where is the wisdom in tenaciously hanging on to our knowledge in circumstances where clearly it is not appropriate? In a world where we have "information super-highways" are we perhaps becoming more and more like the ant, perhaps collecting more and more dead flies?!

So let us beware of the danger of information overload. Too much knowledge without any sense of purpose and direction to our acquisition of information will probably only confuse us and make it easier for us to miss the little snippets—the pearls of great price—that are truly important. We need to choose between what is valuable, and what is not. We need to cultivate wisdom. We surely need to learn to take on board only that which is appropriate, truly of interest and helpful to our task so that we can get on with the job in hand.

Yet it is not only a problem of information overload. It is also about the use, abuse and manipulation of information. For instance, can we truly trust our newspapers and television news broadcasts? Most of what we see, hear or read these days is more opinion than fact. How can we get the truth? I begin to wonder sometimes if the only way is to go to places ourselves, which, with global travel so much more available, is happening. Can we trust those who decide what we should, or should not, know?

So there are three issues: overloading ourselves with knowledge for knowledge's sake; individuals and organisations holding on to knowledge that rightly should be made public and finally the distortion and manipulation of information to hide the truth or to create a false reality. I am left to wonder about the ludicrous situation that we see in companies buying up patents, to the point of even registering patents on the DNA that exists in people who are actually still living! We hear talk of living in an information age with information being seen as a commodity, to buy, to sell, to hoard, to protect. We miss the point completely. Information has value when it is shared. There is information available in our world that, without doubt, would improve the quality of life for millions of people: technologies to replace fossil and nuclear fuels; medicines (natural and chemical) that could control disease. There is information about how power lines, chemical additives in food, pesticides and pollution affect people's health. But the information is systematically withheld, or so it seems, from the public domain.

There is also the gossipy information that the media seems to be full of these days. It titillates people, and somehow it has become so important that we know what celebrities are up to. But what is a celebrity anyway? Often they are people made famous because ..., well, because they have learned how to keep themselves in front of the paparazzi, though some I am sure have lived to regret this. But there is a strange fascination that has been cultivated and, for me, it is a diversion. It takes our focus away from the things that really matter in the world. So a celebrity is seen wearing the same outfit that they did 3 years ago—shock horror! Who cares? I certainly don't. It doesn't matter, it really doesn't, but it becomes somehow newsworthy. I'd rather the media concentrated on looking at who made the clothes, who exploited whom, or was it a genuinely fair trade product?

Every day the media loads us up with useless information that has no value whatsoever. And we are becoming addicted to it. People need to know who's done what with whom, as if it really matters. So what if some celebrity is photographed leaving a club at 2.30am with someone, or the worse for alcohol. It is not newsworthy and it really is not important. Some photographer catches an image of a celebrities breast and wow, this is news, this is full colour spread. But what about the women who are starving, who perhaps have few clothes, who are desperately trying to give milk to their newly born child in some war-torn part of the world? Those are the breasts that matter in our world, the ones that are trying to give life where everything that is happening is trying to take that life away.

Celebrity gossip sums up, in so many ways, what is wrong with the profit driven consumer societies of the world. It obscures what really matters. It is useless information. In truth, it is to be likened to the dead fly.

So I am indebted to the ant (and the fly). And I felt that I should try and pass on what I learned from those few moments of being captivated by the awesome struggle of the ant. I do not recall now what the final outcome was. It somehow does not seem to matter. It was the process, the struggle, the futility, the stupidity that made the impression. Lessons do come in strange forms, and often when you least expect them.

I am left with a recognition that somehow, within the jigsaw of life, there is a need for us to gain a clearer understanding of how knowledge should be used and, even more importantly, the importance of cultivating wisdom. Knowledge is information. Wisdom is about how we use knowledge. The world is full of knowledge. Daily we are seeing our knowledge of the created world extending. I feel that there is some drive within humanity to know, to understand, yet we have surely to go beyond the pursuit of knowledge for commercial gain, or to satisfy some personal desire to feel some kind of titillation. If we do not think for ourselves, and pursue the knowledge that really matters, then we will continue to be fed and to accept what we are told to think. Our minds will be kept away from many of the real issues that matter in the world. Those who control the flow of information in the world have a huge responsibility. Knowledge can be used and it can be abused. Can we trust the information we are given? Do we believe all that we read in the papers? Do we believe all that the politicians tell us? Do we believe all that business leaders tell us about their multi-national operations? Do we?

Our challenge lies in encouraging a greater valuing of and respect for truth, and in cultivating greater wisdom so that the knowledge we gain individually and collectively can be used to lift and ennoble the human condition and safeguard the welfare not only of humanity, but of the planet and the myriad forms of life that we share it with.

◆ ◆ ◆

I have been in love with the sky since birth. And when I could fly,
I wanted to go higher, to enter space and become
a "man of the heights".
During the eight days I spent in space, I realized
that mankind needs height primarily to better know our long-suffering earth, to see
what cannot be seen close up. Not just to love her beauty,
but also to ensure that we do not bring even the slightest harm
to the natural world.[1]

1. Pham Tuan, Vietnamese astronaut. Quoted in *The Home Planet*. KW Kelley (Ed) Addison Wesley Publishing Company. (1998)

10

Trust

Trust is a word that we frequently use and a word which I had not thought deeply about until I had an invitation to facilitate a group discussion on 'Spiritual Living' at a Conference a few years back. It gave me an opportunity to reflect more deeply on this little word because as I was reflecting on what my focus would be, the word 'trust' just kept tapping away in my mind, compelling me to give it attention.

My research began with Webster's dictionary and I found that trust comes from the stem of the word true, *trow* = Icelandic *traust*, trust, confidence; Danish and Swedish *trost*; Greek *trost*, consolation, hope. Well, I had the root of the word, but I didn't feel that I had got much further. I found myself with more questions than answers! What do we trust in? Others? Ourselves? Or something transcending us and them? I found myself finding it harder and harder to keep a firm mental grip on this concept of 'trust' because somehow it seemed to take my thinking into more and more subtle areas. I could think of situations in which I felt trust was present, and those when I distinctly felt it was absent. Yet it remained a mystery.

It seems, however, that trust is crucial, and that we progress through a process of learning to trust ourselves, others and finally the greater 'process' of life. I wonder, do we trust 'in' anything, or are we seeking a trust that has no worldly attachment? Do our difficulties lie in placing our trust in things that are essentially impermanent? We trust the solidity of our house, only to discover subsidence. We trust our investments, only to discover that the markets have fallen suddenly. We trust a friend only to find that they can let us down. Can we trust anything? Is there anything that we can truly rely on?

Perhaps our problem lies in that we often think of trusting something in the sense of trusting that something will happen; and more often than not this is about trusting that something will work out the way that we would like it to. The problem is we have no control over events. The unexpected all too often happens.

So in a way this is a kind of false trust. Trust in something has to be more than wanting something to work out as we want it to.

So what can we trust in? My pondering in preparing for that Conference led me to believe that we can only truly find trust when we trust without strings, without conditions, without preconceived hopes and expectations. Trust is often bound up with desire, our desire for something and we saw what can happen with desire in the parable of the Wish-Fulfilling Tree!

It seems that at risk of sounding vague and nebulous we have to trust what is and what will be; to trust ourselves, to trust the moment, to trust the process of life. Trust is so much more than trusting someone to do something, or to be a particular way—the way that we want them to be. Yes, it may well include this, but that is not the whole story.

It seems that to live spiritually is to live in trust, and this a trust focussed not in or towards anything in particular, but perhaps rather in everything. In other words, we cannot trust a part of creation—ourselves, others or whatever—in isolation. We either trust the whole, or nothing. Trust is an all-embracing quality that extends everywhere if it is to be real. And the greater process may not bring us what we want or desire or hope for. We cannot guarantee that things will work out as we wish them to. We cannot trust that. We can hope. We can believe. But we cannot be certain.

This seems to be at the crux of it. What is the nature of trust in a world that is fundamentally uncertain? We have to trust the larger picture, the larger process, that there is a wisdom and an intelligence, and a purpose, working out through ourselves and the world around us. We come back to the metaphor of the jigsaw. Can we trust that this is how life is? Can we trust that there is a bigger picture that is being created and of which we are each a small part? We want to do what *we* want, full of our own self-interest and self-importance and by so doing we do not see the bigger picture. We place our trust in the impermanent, the changing and shifting shapes, colours and patterns of the pieces. And then we are horrified to find that they change in ways that we had not wanted or anticipated. Such changes can often be those linked to losing that which we have become attached to. But the larger picture demands to appear and life will give us the experiences we need in order to eventually learn that the world is more than what we want, more than our own personal self-interest and desires. But it can feel so very difficult to trust that greater process because often our trust is linked to comfort, to feeling in control, to things being predictable, and to the hope that we will be safe. Life, however, is not like that.

What blocks us from trusting is fear; fear of the unknown, fear of being hurt, fear of being isolated, fear of being fearful. So we invest our trust in the things we can grasp, the things that we believe will make us secure and happy. But they will not do this for us. The more we attach ourselves to the impermanent, the more we will find that it is not so satisfying after all. It is unpredictable and we find that it can be untrustworthy in the sense that we cannot trust it to give us what we think we want. Or it does give us what we want, but it gives us other experiences as well that are not as we would wish. We desire fame, but if we gain it then whilst we may be free to travel the world with our money, we may lose our freedom to be able to walk to the corner shop because of media attention. We desire money to take away our worries, but we end up worrying about what to do with our money, or how to ensure others do not take it away from us, or how to make even more!

A deeper trust

Where does trust reside? I think it has to be a heart quality. It is surely born out of a depth of knowing that is not of the intellect, but emanates from a sense of connection with a greater process, which we call evolutionary growth, and which is driven, perhaps, by some unforeseen force, maybe by Love.

> "What is the treasure of the heart? Not only benevolence, not only compassion ... but consonance with the Cosmic Consciousness when the heart, besides its own rhythm, even partakes in the cosmic rhythm. Such a heart can be trusted; it possesses straight-knowledge, it feels and knows, and as a manifested link with the Higher World it expresses the indisputable."[1]

Roberto Assagioli, founder of *Psychosynthesis*, used to work with triangular relationships between qualities, seeking to uncover how opposites may be transcended in a higher quality. I used an example of this when we looked at unanimity in an earlier chapter. We can think about this in relation to trust. What undermines trust? It seems to me that here we need to think about uncertainty, and its opposite—certainty. Life is uncertain, but we try to make it certain and secure. And when life is certain we sometimes feel the need to break out of its constraints. Reality is that life is an uncertain experience yet I would suggest that it contains the certainty of growth—eventually. We can take a very passive attitude towards the certainty and uncertainty of life, simply accepting what will be in a rather fatalistic and 'dumped on' approach. This is certainly not very cre-

1. *Hierarchy*, Sutra 106. From the Agni Yoga series. Agni Yoga Society Inc., USA

ative. Or we can think more positively and consider trust as marking a point in the relational triangle that transcends the certainty/uncertainty divide, a place from which we can become active participants in the greater process of life. The following relationship therefore emerges:

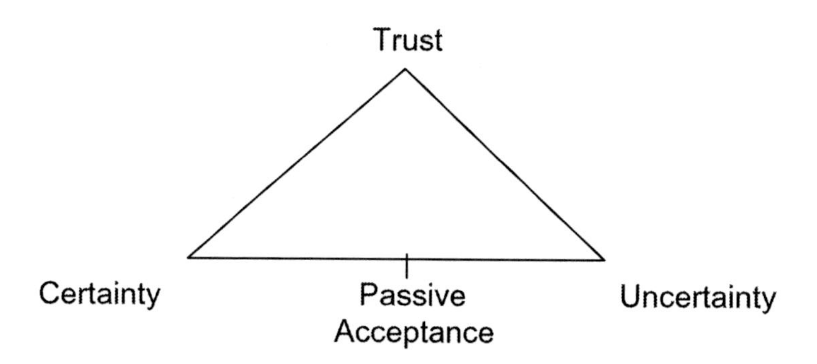

Can we find our own path to the point of trust, where both certainty and uncertainty can be embraced, and in a sense transcended? A place where we can let go of worldly attachment, where we are no longer constantly seeking to create certainty? In our hectic world where security and happiness is promised if we buy this, or do that, this appears to be one of our greatest challenges.

We seem to have to recognise and trust that we are where we are for a good reason, for a purpose. It does not help us if we spend our time making excuses or looking for ways out, or feeling helpless. Life is the great teacher and experience provides us with our lessons. We need surely to learn to trust ourselves and the situations in which we find ourselves. Having said that there are also times when it is right to look for change. Where circumstances are life-threatening, where we are the target for abuse, we may well seek greater immediate security. Again, we have to trust our inner sense of what is the right thing to do.

> "Trust thyself: every heart vibrates to that iron string. Accept the place the divine providence has found for you, the society of your contemporaries, the connection of events. Great men have always done so, and confided themselves childlike to the genius of their age, betraying their perception that the absolutely trustworthy was seated at their heart, working through their hands, predominating in all their being."[2]

---

2. R W Emerson; Essay entitled 'Self-Reliance', *Collected Works*

There is power in trust and a potential to achieve great things on the world stage.

Trust in a therapeutic context

I am reminded of Carl Rogers, the American psychologist and founder of the Person Centred Approach to counselling and psychotherapy. He had some powerful ideas about trust and his perspective on therapy has been a major factor in influencing my development in counselling and therapeutic engagement. He emphasised the essential trustworthiness of the person. He believed deeply that, in a climate of relational genuineness, empathy and unconditional positive regard, a growth process could be trusted to take place.

> "... organisms are always seeking, always initiating, always 'up to something'. There is one central source of energy in the human organism. This source is a trustworthy function of the whole system rather than of some portion of it; it is most simply conceptualised as a tendency toward fulfilment, toward actualisation, involving not only the maintenance but also the enhancement of the organism."[3]

He expressed a belief, and one with which I agree, that in a therapeutic situation where he could touch the deeper, transcendent aspect of his own nature, he could trust himself to speak or act in ways that were growth enhancing for the other person. He wrote of how, where there was a sense of deep connection with his client and with his own inner core of being then simply his presence became "releasing and helpful to the other". He wrote of this that:

> "There is nothing I can do to force this experience, but when I can relax and be close to the transcendental core of me, then I may behave in strange and impulsive ways in the relationship, ways which I cannot justify rationally, and which have nothing to do with my thought processes. But these strange behaviours turn out to be right, in some odd way: it seems that my inner spirit has reached out and touched the inner spirit of the other. Our relationship transcends itself and becomes a part of something larger. Profound growth and healing and energy are present."[4]

Trust seems to be one of the magical and transformative ingredients in the healing process, whether we are talking about physical or psychological healing.

---

3. Rogers, C (1980) *A Way of Being,*. Houghton Mifflin Company, Boston. p. 123
4. Rogers, C (1980) *A Way of Being*. Houghton Mifflin Company, Boston p. 129

When we are working with people, supporting them, being companions with them in their struggles, there is a need to validate people's feelings. We so often want to rush in and make someone better. It is difficult to trust their process and to simply hold them where they are, allowing them their own space to move around in freely.

This was emphasised for me when working with a group of health-care professionals. We were exploring working with someone who had relapsed in seeking to break an alcohol dependency. They all rushed in (in role play) and tried to make the person feel better by coming up with good positive ideas to help them unravel what had happened and feel better about it. Such responses, of course, do have their place. We all do this. We are human, we want to help. We want to heal. We want to ease discomfort. And we also want to stem the flow of our own discomfort! But for me there is something about those periods of psychologically and emotionally holding someone, of validating their feelings (and this is painful for us too) because the climate of relationship has healing potential. We sit with them, maintaining a relational attitude of empathy and warm acceptance towards this other person, being genuine in our attitude and responses, and maintaining our trust in the process that is occurring for our client, and indeed the relational process that we are party to.

Consider the following:

> "Enter thy brothers heart and see his woe. Then speak. Let the words spoken convey to him the potent force he needs to loose his chains. Yet loose them not thyself. Thine is the work to speak with understanding. The force received by him will aid him in his work."[5]

Can I truly learn to trust an individual person's process of growth? Not easy when you have that person in front of you pouring out problems and feelings that are just so overwhelming, with often so much despair and hopelessness. But I then find myself asking myself, "OK, so what is this really all about?". It is my belief that the more we live from the core of our being, from what I would call the Soul and which others might describe as a state of fuller functionality, the less we live by chance. People hit crisis for a reason, and I was immensely heartened to come across a book by Robin Norwood which I could quote at great length from. She is a psychologist who has embraced the spiritual and the notion of a greater evolutionary process of conscious growth in her work, and presents them

---

5. Bailey, AA, (1934) *A Treatise on White Magic*, Lucis Press Ltd., New York and London. p. 320

in an extremely powerful way within the therapeutic context and particularly in relation to family systems. I will include one short passage:

> "When we attempt to save another from an illness or problem, we may inadvertently be interfering with that person's reason for incarnation—the enlightenment he or she is pursuing under the direction of the soul. This is especially challenging when the person we want to help is our child. Honour another's path, another's karma. If watching is too painful, we need to get help for ourselves to better handle our own suffering."[6]

Being with those who suffer is a privilege for us, but we must be open to their suffering. *How can they feel our love if we do not feel their pain?*

In interacting with people I have learned that my greatest resource is myself: How I am with people. And this is rooted in how I am with myself. We come back to trust again. Can I trust myself to do the best I can? Can I trust that that will be good enough? Can I trust that I shall say the 'right' things to people? Who knows? I don't know what the 'right' things are, necessarily. I merely offer my presence in relationship with someone, and trust that if the Soul nature can become present through this relationship, then healing and growth may occur—where this is in line with the Soul's intent. My hope is that people will be touched in a way that will help them to "loose their own chains" and become the person that they in essence truly are, or to take a more psychological line, realise the potential that they have.

To just add in a further angle on trust, and one which I will not overly dwell on here, but many spiritual traditions speak of a process of initiation. This is the idea that we undertake a series of initiations that mark expansions of consciousness in which the individual becomes more fully conscious of the Soul and of the Plan and Purpose of God. It seems to me that to be initiated is surely to be entrusted; entrusted with knowledge but also entrusted, perhaps, with greater capacity to wield certain forces and energies to serve a greater spiritual purpose. And as one achieves higher and higher initiations then this trust grows, and with it the responsibility too. Not only do we become trusted with certain energies, but also knowledge and responsibility to express what we now know in a life of selfless service. Perhaps in the jigsaw of life there are some pieces that find their true shape, colour and pattern sooner and become as magnets, drawing others to them in some mysterious way in order for areas of the final image to take form.

---

6. Norwood, R (1995) *Why Me? Why This? Why Now?*, Arrow, London. p. 215

For me, the presence of trust in relationship with another is of great therapeutic value. It is also an expression of my caring and sensitivity to their process of growth. I believe that trust is another key piece in the jigsaw of life. To live in an atmosphere of distrust can breed fear, separateness and exploitation. And it can be extremely stressful. Is this how we wish to see the future? The more we encourage the capacity to trust, the more trustworthy people may become. I do not mean a blind trust, we know that there are people who threaten the well-being of others and can arguably be trusted to only behave in that way. But I kind of feel we have got things out of proportion with fear dominating trust and my hope is that we will find ways in the future of addressing this. And somewhere within this process is the need for the Soul nature, the deeper spiritual sense of self, to emerge more fully into conscious awareness. Then, perhaps, we will see a world in which trustworthiness will become more apparent. Then there will be a trust that is centred in that which is responsive to the permanence and certainty of God's Plan and Purpose for humanity.

Your Time Is Now

A cold wind blows across the face of the one who stands
Looking out into the void beyond the edge of life.
An abysmal moment.
No movement. All is still. A distant call ..., then silence.

The past stretches behind, lost in the darkness of forgotten memories.
And the future? Hidden from sight.
Does it really exist? Who brings it into being anyway?
You? Me? Perhaps we never do.
It remains, eternally, that one step ahead of us.

But the only step now is in the present.
A step that leads ... where? Oblivion? Or Glory?

The waves crash upon the jagged rocks below,
A thunderous tumult that beckons the unwary.
Somewhere above, the sun, obscured from sight.
But you are all alone now. One step, just one step away from your Self,
Tormenting your indecision.

Standing still you remain locked on a wave of probability.
Waiting, always waiting. Inert consciousness unmoved.
Collapse it. Step forward. Are you a rock or a seagull? Take that step.

The darkness thickens and the eternal moment arrives.
Step into Being and Know your Time is Now.

11

Stress: A Holistic Perspective

> *"Only in the stress of circumstances can the full power of the soul be evoked. Such is the law"*[1]

We hear the word *stress* mentioned increasingly these days. A great deal of ill health is linked with high levels of stress. It may stem from home life: the children, relationship difficulties with a partner or family members, a move of house, an elderly relative to care for. Or it may originate in the workplace: overwork, inadequate working conditions, unrealistic deadlines, lack of support/supervision. Stress may originate through change, internal and external, through movements that occur within our own natures. Change involves uncertainty which can cause stress. We seem conditioned into experiencing a need to know what is going to lie ahead of us, even though we know that life is uncertain as highlighted in the previous chapter. Indeed, we might say that there is a degree of linkage between trust and stress, with stress more likely where there is an inability to trust what is happening, or trust is invested in that which is uncertain, unreliable and impermanent.

Stress is defined in *Chambers* dictionary as: a "constraining influence" and "system of forces applied to a body". I would argue that *stress is the effect of having to do something that you do not want to do.* In a work situation you are given a task that you feel inadequate to cope with, or which you simply don't want to be party to but are unable to say no. It may even be something you enjoy doing, but not for twelve hours a day. Stress is an effect of an imposition that is overloading your capacity to fulfil what is demanded of you and preventing you from doing what you really want to do. You are constrained by external forces which impinge on your personal time and space, on the activity of your personal energy field.

---

1. Bailey, A, (1944) *Discipleship in the New Age, Vol. 1*, Lucis Press Ltd., London. p. 181

Stress is also an indicator of boundaries. We each have weak points that tell us we are under stress: backache, headache, irritability, insomnia, lack of motivation. We can build a picture of what we are comfortable with in our sphere of energy, and what we find stressful. We can sense which external influences encourage a sense of fulfilment, and which diminish and disempower us. It puts us in a strong position to choose a path through life. It is certainly rare to feel completely stress free. Indeed, it could be argued that a lack of stress might be an unhealthy condition, seen from a more holistic perspective.

So far, our theme here has been what could be called 'worldly stress'. We need to ask, though, whether there is a deeper cause, a deeper dissatisfaction with life and the experience it gives us, a stress that emanates from somewhere deeper within our natures, a stress that develops from the presence of a certain dissonance between the human personality and the spiritual nature that may be seeking expression?

Are we stressed by clinging to old forms and habits, old ways of being and of functioning that block the Soul nature from finding expression? Are we experiencing an increasing inner tension and flow of energy that has no outlet? We might call this 'holistic' or 'esoteric' stress. 'Esoteric' is used to denote *inner*, suggesting a quality of stress arising out of inner processes that are concerned with individual spiritual growth and development. For the person seeking to cultivate spiritual sensitivity, stress has a positive role to play in invoking the life and energy of the Soul.

It is not always recognised among those working in stress management that human spiritual sensitivity produces stress when it is obstructed from finding expression. The more we are attuned to the inner, deeper impulses that seek expression through us, though we may continue to experience a certain 'spiritual tension', we will not feel the same degree of stress. Why? Because we are now living and functioning in tune with the inner impulse, with the Soul nature seeking expression. In terms of the jigsaw metaphor, we are beginning to realise the colour, shape and pattern that is intended for us and beginning to find something of our place in that bigger picture. There is, I think, an overarching will to establish that final picture and the more we block this, the more we will feel stress. As we attune to this deeper, inner process we adjust in some mysterious way to something that has a certain quality of 'rightness' about it.

Until we find that rightness, however, we will be seeking ways of relieving the stress. We might use substances, develop all kinds of habits and addictions to distract us or to take away deep discomfort and stress. Indeed, in some instances this stress will be linked to the concerns and compassion we feel towards the plight of others, and to the world in general. The motivation to relieve such stress in ourselves at witnessing the suffering of others can induce a tremendous will to serve. This, of course, leaves us with the question as to whether we serve to meet the needs of others, or to reduce our own stress and pain at witnessing another's suffering. Motives can start to get a little mixed.

Soul influence stimulates refined sensitivity within our individual natures. How can we then look at the suffering of humanity with open and sensitive hearts, whilst finding ourselves pushed into behaviours that to some degree collude with separateness and contribute to human suffering, without feeling stress? We drive our cars, and pollute the air. We invest our money and find it invested in turn in unethical activities. We buy the highly packaged food from the superstores and signal our agreement with their policies. We eat food that originates from the cash crops that the poorer countries have to grow to pay the interest on their debts to the world banking system, whilst the people producing the crops go hungry. We buy and wear the cheap and/or fashionable clothes that turn out to be made in a sweatshop where people are overworked in dangerous environments with little health and safety considerations

Sensitivity to the Soul is a cause of stress. The Soul often demands something of the separative personality which it simply does not want to do. The treading of a spiritual path is the greatest initiator of stress. It is a guaranteed outcome! Make a commitment towards spiritual growth and we will experience stress as we have never done before! And the answer does not lie in trying to take it away or changing what we want to do, unless we change towards greater responsiveness to the Soul's intent. Even then, we will still find stressful circumstances arising, but we will at least be using them to re-create ourselves and to expand our consciousness.

The stress that arises out of seeking to find a spiritual way of being in the world produces all kinds of effects on our natures. This may take physical form in disease; it can induce an imbalance in nervous energy. The feeling nature can be thrown severely off-balance. It can produce a disorientation in our thought life, generating chaotic thought processes that distort our interpretation of what occurs in and around us. It is wise to be forewarned, Soul energy is powerful and if we are unable to maintain a steady alignment, the downflow of heightened vibration will throw us about emotionally, mentally and physically, generating tremendous stresses on the system.

We need to put aside the thought of getting rid of stress, or of seeing it as something to avoid. It is inevitable. We can use it to learn about our own capacities, recognising the signals we each receive when stress is building up. Then we can look for constructive recreation, to give our systems time out to re-energise, for instance through contact with nature. Perhaps one of the causes of so much stress in our inner-city areas is due, in part, to lack of contact with, and an encouraged appreciation of, nature and the natural world? Meditation and prayer can bring us freshness, inspiration and renewal. Simple communion with fellow human beings can also help. What is crucial is that we act in a positive, not a passive, manner when we seek constructive recreation.

If stress is the effect of a build up of force that is blocked from expression by our natures and personalities, then we need to seek ways of expressing it that are rewarding and creative, that lift our spirits. We might want to change our environment for a while, seeking resonance with the sounds of nature rather than the all too familiar noise of car engines, tyres on roads, and electronic clicks and buzzes. And, of course, there is the 'noise' that is inaudible, the insidious expansion of electromagnetic fields and radiation that pervades the 'developed' and 'developing' worlds. Perhaps part of our stress stems from not having enough range of genuinely natural vibration around us.

We need to be clear as to whether:

- our stress is simply the effect of overload or of trying to manage/cope with too much at any one time
- different elements of our personality nature pulling in different directions, with the need for us to work on and strengthen our integration
- personal attitudes and ingrained patterns of behaviour that are blocking Soul impulse
- we are in the process of adjusting to the Soul as it imposes a new rhythm on our lives and a more inclusive awareness on our consciousness

Dealing with stress can only be done on an individual basis. There is no universal panacea. Methods will differ from person to person, and the cause of the stress and the degree of sensitivity to the Soul nature will be a crucial factor to consider. In many respects, stress from a Soul perspective appears to take an opposite view to the generally accepted one which regards stress as something to manage, offset and, where possible, remove completely. The Soul-centred individual acknowledges the need for stress and learns to ride it, rather like a surfer rides a wave. He or she will use it to discover inner points of tension from which they can draw inspiration in their work.

Are we, as evolving pieces within the jigsaw of life, seeking to consciously fulfil the purpose God has for us which the Soul nature is responsive to, and to be able to respond in such a way that we are able to become the piece that is envisioned in the final picture? Or are we to 'do our own thing', continually ignoring the inner promptings, and the true, spiritual teachings and reminders that others have given, and continue to give us, in response to the world situation? If we genuinely choose to respond to the Soul nature then we will experience the stress of having to let go of certain separative and self-centred ways of being and behaving. Stress of this kind, we might say, is a natural experience with the process of finding our shape, colour, pattern and place within the jigsaw of life. And when we achieve this 'letting go' we will become as the lame boy in the parable of the Wish-Fulfilling Tree. We will forget to wish for ourselves. Our attention will be on the plight of others. We will open ourselves to greater compassion. And we will detach ourselves from the universal processes that bind us through our desires to worldly experience.

The Soul is calling us. We ignore her at our peril. If we do ignore her she is sure to ensure that the stress of circumstances will arise that will force us to draw her more fully into our lives. And if we respond to her calling, similar circumstances will still arise to enable us to evoke her power even more fully. "Only in the stress of circumstances can the full power of the soul be evoked. Such is the law."

12

Person-centred Psychology: From the Personal ...[1]

The latter half of the 20th century has seen an enormous growth in what are loosely described as 'the talking therapies': counselling and psychotherapy in particular. In this chapter and the following chapters the term therapist is used to mean both counsellors and psychotherapists. Many people are entering training programmes, attracted for a whole host of reasons, some more to do with their own personal development, others concerned more with acting on an urge to help others, and still others simply seeking a new career and a way of earning money. We also witness masses of people going 'into therapy' and this again for a whole host of reasons. Perhaps it is a requirement of a training course. For others it will be to make sense of something, or to resolve a discomfort. Still others may seek therapy as a way of enhancing their functioning as a person. Then there is the fact that there are so many different schools of therapy, different theories and ideas that the client has to choose between. So much creativity has been focused in recent decades in creating models for helping and offering therapeutic support.

One of the core theories of therapy was that established by Carl Rogers, what he termed 'client-centred therapy' or 'the person-centred approach' (PCA). His ideas evolved as his own life proceeded and led him to suggest later in his life:

> "I hypothesis that there is a formative directional tendency in the universe, which can be traced and observed in stellar space, in crystals, in micro-organisms, in more complex organic life, and in human beings. There is an evolu-

1. This chapter and the next is largely taken from two articles: 'From Person to Transperson-centredness: A Future Trend?', first published in *The Person Centred Journal* (1997) Vol. 4, 1. and 'Carl Rogers' Client-Centred Therapy: An Esoteric Perspective', first published in *The Beacon* (1999) July/August. Lucis Press Ltd., London and New York.

> tionary tendency towards greater order, greater complexity, greater inter relatedness. In humankind, this tendency exhibits itself as the individual moves from a single cell origin to complex organic functioning, to knowing and sensing below the level of consciousness, to a conscious awareness of the organism and the external world to a transcendent awareness of the harmony and unity of the cosmic system, including mankind."[2]

In terms of the potential of the individual, he suggested that there exists within the person what he termed as an 'actualising-tendency', a movement that constantly seeks to take us towards fuller functioning. He wrote that:

> "Individuals have within themselves vast resources for self understanding and for altering their self-concepts, basic attitudes, and self directed behaviour; these resources can be tapped if a definable climate of facilitative psychological attitude can be provided."[3]

He suggested that certain conditions, when present in a relationship, encouraged growth. He described them in a paper in 1957 as being the 'Necessary and sufficient conditions for therapeutic change'.

> "1. Two persons are in psychological contact
> 2. The first, whom we shall term the client, is in a state of incongruence, being vulnerable or anxious
> 3. The second person, whom we shall term the therapist, is congruent or integrated in the relationship
> 4. The therapist experiences unconditional positive regard for the client
> 5. The therapist experiences an empathic understanding of the client's internal frame of reference and endeavours to communicate this experience to the client
> 6. The communication to the client of the therapist's empathic understanding and unconditional positive regard is to a minimal degree achieved"[4]

Rogers was presenting a view that human beings are essentially relational and that when we enter into genuine, unconditional, empathic relationships, growth occurs. He also affirmed that the individual can be trusted to have the wisdom to know what their own needs and direction are.

---

2. Rogers, C (1980) A Way of Being. Houghton Mifflin, p.133
3. Rogers, C (1980) A Way of Being. Houghton Mifflin, p.115
4. Rogers, C (1957) 'The necessary and sufficient conditions of therapeutic personality change. Journal of Consulting Psychology 21. p. 96

I trained in the Person Centred Approach in the early 1990s and what I came away with from training in this approach was a sense of my need to be able to be open in my communication and interaction with other people. I needed to first of all be aware of what I was experiencing and be able to interpret it accurately in terms of its meaning. I also needed to be sensitive to the experiencing of the other person and to create, through our relationship, a certain quality of interaction which would encourage the other person perhaps take risks in their struggle to experience themselves in a more accurate way and thereby gain fresh insight into their own nature and way of being.

As I absorbed myself more and more in this approach I found myself naturally wondering how the spiritual side of my world-view could be connected to what started out for Rogers as very much a humanistic approach. The result of these musings was a paper which I have edited and reproduce here in this and the next chapter, and to which I have added further ideas that I have developed more recently. It reflects my own process of seeking a bridge between the strictly psychotherapeutic PCA and the possibility of becoming more than a person in the strictly human sense. Rogers wrote and spoke a great deal on the theme of 'becoming a person'. For me we need to re-evaluate what we mean by a person. Are we strictly limited to the flesh and blood experience of a physical form driven by chemical stimulus to reach as far as possible a state of fully functionality? Or are we more? Is there scope for reaching what I call here, 'Transperson-centredness'?

From Person to Transperson-centredness: A Future Trend?

The essential idea behind the Person Centred Approach (PCA) is a belief that all organisms are evolving; that given appropriate conditions they can grow to realise fuller potential. Rogers suggests that we each have a tendency, latent or realised, towards growth—a *formative tendency*. For Rogers, this formative tendency was not limited to human beings, it extended to all living forms from crystals to …, well, perhaps even the cosmos itself? His vision fills me with a sense of being a participant in a vast, living process of unfoldment, carrying us all God knows where. We might equally think of this tendency as an evolutionary impulse, a force that operates throughout creation to bring about some form of fulfilment – a condition of *wholeness*. It is all-pervasive, transcendent and immanent, constantly urging movement in a growth-enhancing direction.

When, in a relationship, an individual is fortunate enough to experience the three core conditions—empathy, congruence and unconditional positive regard—then they are more likely to also experience a triggering of what he termed the 'actualising tendency', a movement that takes them towards actualising their potential as a more complete person. Perhaps triggering is not the best word to use for the actualising tendency is ever present, urging the individual to maximise the potential in any situation to bring a satisfying experience. Maybe it is more that when the relational conditions are right then the direction that the actualising tendencies urges the individual in is more likely to be constructive, hence the phrase 'constructive personality change' to which Rogers refers. The core conditions need to be *experienced* in at least some measure by the client. Rogers' emphasises this:

> There is ... one condition which must exist in the client. Unless the attitudes I have been describing [core conditions] have been to some degree communicated to the client, and perceived by him [or her], they do not exist in his [or her] perceptual world and thus cannot be effective.[5]

Elsewhere Rogers writes of this in terms of the client seeing that he "is psychologically received just as he is by the therapist".[6] The PCA perspective argues that this can encourage growth.

The influence of the actualising tendency enables an individual to move from where they are in themselves, psychologically speaking, towards a state in which they realise greater freedom from the conditioning effects of past experience. Rogers highlights a belief

> that there is one central source of energy in the human organism; that it is a trustworthy function of the whole organism rather than of some portion of it; and that it is perhaps best conceptualised as a tendency toward fulfilment, towards actualisation, not only toward the maintenance but also toward the enhancement of the organism.[7]

Just how the 'actualising tendency' operates is a mystery. We are left to wonder whether it has its roots in the physical organism itself, whether it acts from more subtle realms of thought and feeling, whether it is actually innate in consciousness itself or, if it is something that finds expression through the whole of

---

5. Rogers, C and Stevens, B (1973) Person to Person. Souvenir Press, London. p. 96,
6. Rogers, C,((1961) On Becoming a Person. Constable and Co. Ltd., London. p. 131
7. Rogers, C (1978) Carl Rogers on Personal Power. Constable and Co. Ltd., London. pp. 242-3

creation. Perhaps it is simply a fundamental process of nature of which we each are but a part.

A question of growth

If we accept the hypothesis that an experiencing of the core-conditions ensures that the actualising tendency encourages growthful and constructive personality change, we are then left with the questions:

What do we mean by growth in terms of the person-centred approach?

Where are we growing from?

What are we growing towards?

These are crucial questions and cause us to turn to the theoretical roots of person-centredness. The whole theory is, in my view, 'growth-oriented' and rests firmly on the hypothesis that all organisms have a growth potential. However this potential is distorted and blocked by how we react to our experiences and how this, in turn, shapes our self-concept. The tendency to grow, to seek our fullest potential as an effective and sensitive organism continues in spite of what we experience, but it will take us along a different growth curve. Rogers suggests that we experience 'conditions of worth', which encourage us to establish a negative self-concept, conditioned by how others react to us. Negative conditions of worth evolve out of the put-downs of childhood and the lack of prizing or praise from parents and significant others which most of us have experienced, or the conditional praise often given to children. Or they may be linked with far more traumatic experiencing that, in extreme cases, can cause the creation of dissociated states within our structure of self, places we escape into, or places that may simply be extremely damaged and vulnerable that we then try to keep away from.

The person-centred approach seeks to enable the individual to come to terms with their conditional self and to develop in ways that free them from past conditioning. This often means the client discovers a need to recognise the nature of his or her self-concept, to experience the pain that is associated with this, fully and deeply, and to then realise that it is not how they truly are or have to be, but how life-experience has formed or conditioned them to be on their individual path of becoming. From this recognition, and through a re-integration into the conscious nature of the thoughts and feelings associated with the conditioning, the individual can move away from a negative and somewhat incongruent self-concept towards one that is far more positive and realistic.

In person-centred terminology we are concerned here with a shift in a person's 'locus of evaluation'—the psychological focus from which individuals evaluate themselves. An aim of a person-centred practitioner is to create the relational climate based on person-centred principles such that it enables the client's self-evaluation to become less centred around the opinions of others, and more upon their own thoughts and feelings which are experienced with increasing accuracy and without being distorted by conditions of worth. Rogers writes:

> Another trend which is evident in this process of becoming a person relates to the source or locus of choices and decisions, or evaluative judgements. The individual increasingly comes to feel that this locus of evaluation lies within himself. Less and less does he look to others for approval or disapproval; for standards to live by; for decisions and choices. He recognises that it rests within himself to choose; that the only question which matters is, "Am I living in a way which is deeply satisfying to me, and which truly expresses me?". This I think is perhaps *the* most important question for the creative individual.[8]

Facing up to our conditionality

In terms of our organismic growth (or lack of growth) we find that as we face up to our conditions of worth in a climate founded upon the core conditions, we begin to move towards the place where unconditionality might have taken us. We can never be sure of this, and it would be foolish to make dogmatic assertions in such areas of great subjectivity. The working of the actualising tendency helps us to move towards greater wholeness. Conditionality means we take a less than direct route, and perhaps it will be a different quality of wholeness that will result. I am reminded of chaos theory and of how small initial changes can lead to large differences in outcome. Maybe it is not the form but the quality of our wholeness that matters with common elements of beauty, harmony and integrity the key. Maybe we should define wholeness less in terms of a state or condition, and more in terms of quality and presence.

In a world that seems to encourage alienation, judgemental attitudes and a lack of acceptance of people's diversity and difference, Rogers' core conditions can help us to acknowledge how we really feel about ourselves and the world. By expressing where we are, and what we feel, we are challenged to choose whether to continue to accept our conditioned selves in the light of our new self-knowledge, or embrace what this light reveals and thereby initiate a cycle of re-creation out of which a new self-concept can flourish, and a new way of being emerge.

Somehow, a re-integration of the feelings and thoughts associated with conditioning in an atmosphere of trust, realism and empathic understanding, enables a person to experience what Rogers termed 'moments of movement' Although

---

8. Rogers, C ((1961) On Becoming a Person. Constable and Co. Ltd., London. p. 119

these are hard to describe, they seem to be moments of deep connection in which a triangle of open communication is formed between the client, the therapist and some particular feeling or thought. As a therapist in that moment I feel an expansion within my being. I am reaching out, indeed, I am tempted to say I am 'reached out' to the other person who is my client, as if something more real in me is at work. All barriers are down, a kind of unity ensues based on an unseen yet poignantly present deep connection. The client embraces some aspect of themselves that was previously unknown, or sees something of themselves in a fresh light. No longer blocked from an area of experience, a moment of both intra-and inter-personal connection and communication occurs. The client is freed up to move on, no longer held back by past conditioning. Such moments play a key role in the therapeutic process of *becoming a person*. For me, they have a truly spiritual tone. Why do I say this? Because surely the spiritual state is one of wholeness and fullness. And I acknowledge that from a humanistic perspective such experiencing is quite possible without any notions of a God. Whilst for others it will be seen to be linked to moving closer to the God immanent within us (the Soul nature) or God transcendent.

Where does this movement lead us in terms of the individuals experience and sensitivity? For me, it is concerned with greater capacity for what I would term 'authentic living'. It means the person becomes less fragmented, and therefore more joined up within themselves, more open to the fullness of their own experiencing. There is less distortion or dissonance between the way they see themselves and the way that they truly are. There is also greater capacity for unconditional positive regard towards themselves and others. And there is also greater capacity for empathy. The person is more open and sensitive to themselves and the world around them, and the scope for this sensitivity widens or extends to farther horizons.

Rogers wrote of 'the fully functioning person' – a way of being that is *open* to experience, that exhibits behaviour appropriate to circumstances. There is a sense of moving from a state of fixity towards one more characterised by flow and movement. The individual is sensitive to his whole organism and is able to trust its promptings. It denotes a condition of integrity and of identification with one's whole self. As Rogers wrote:

> It appears that the person who is psychologically free moves in the direction of becoming a more fully functioning person.[9]

To be psychologically free is to be open to the reality of human experience without being limited by conditioning. It was Rogers' belief that through the application of PCA in therapy, psychological freedom could be encouraged. For me, it can be summed up as: *we grow towards a new way of being that frees us to a fuller experience and expression of our own wholeness.*

This is a very optimistic perspective and one that gives us hope. I would like to next explore the nature of the potential that we can discover within us through the growth process and where a triggering of the actualising tendency might lead us. Could this be into realms beyond those of accepted psychological growth? Are there greater potentials that we can discover and actualise that may appear to be from other dimensions of being, or of aspects of our nature which many would describe as spiritual?

It seems that this is all significant and it might be helpful to take a more spiritual perspective on this phenomenon. But where do we begin? A view that we might take would be that human beings are essentially part of a created universe that experiences an urge towards growth. We are participants in an evolutionary process that tends to take everything towards an expansion of consciousness and greater, intelligent and compassionate functioning. As individual persons we are perhaps like sparks of consciousness put down into matter, and striving to regain contact and absorption once more into the Soul nature, the flame of the spirit. Not everyone is conscious of this, but many are, and are seeking ways of connecting with their deeper selves.

False-conditioning

I would suggest that as well as being exposed to the 'negative conditioning', as described previously, we are also affected by what is, in reality, a kind of 'false conditioning', in the sense of being conditioned in ways that are at odds with our inner Soul natures. Such false conditioning may appear and feel positive, for instance, to go out and be successful, to strive for more material possessions and to get to the top. All very sound and positive stuff, but is this what our spiritual natures demand of us? Where is the compassion and the loving kindness? Where are the feelings of community? Where is the sensitivity to the needs of others and to our own inner, aesthetic qualities? I think we need to widen the scope of 'conditions of worth' and to differentiate between negative and false conditioning.

---

9. Kirschenbaum, H and Henderson, VL (Eds) (1990) The Carl Rogers Reader. Constable and Co. Ltd., London. p. 419

Perhaps we should dwell a little more on this distinction. Negative conditions of worth provide us with a secondary valuing system, distorting our self-concept and affecting our functioning capacity. We become unable to act freely and openly in the world. We lose our childlike wonder and openness to every new experience. We are having to adapt ourselves—the way we behave, the way we present ourselves to the world—to meet the needs of others, to feel the warm acceptance that we crave from those around us. Our sense of self-worth and self-valuing becomes intimately linked to the way others see us, or what they want from us.

False conditioning, in contrast, provides an overlay that is based on a fundamental belief that human beings are separate from each other and that there is no deeper sense of self. It is likely to emphasise competitiveness and selfishness, compounding the distortion of negative conditioning. This may then be exacerbated by Darwinian-like urges from our organism to succeed and survive in a hostile and threatening world with the imperative that we have to be the fittest to survive. If we believe we are separate from each other at a fundamental level, and that there is no deeper connection possible, nor any greater purpose seeking expression through our individual and collective lives, then we live according to a set of rules that are, in effect, what some would term 'laws of the jungle', except in the jungle there is an eco-system that if left alone will thrive. With humanity, the sense of separateness is actually one of the core threats to its survival. Human beings have the ability to choose their behaviours, but where this ability to choose is severely affected by the negative and then false conditioning described above, individual interests, or those of a group of people, will be placed before the needs of others or of humanity as a whole. And this we see working out in the world today, whether it is the exploitative organisation or dictatorial, totalitarian regime.

Is all this an inevitable facet of the growth cycle? This, for me, is where it gets particularly interesting, and challenging. To reach the spiritual component within us we have to move through a phase of getting in touch with our humanness, a humanness that has been largely shaped through the false, worldly conditioning that tells us we are separate beings in an unconnected world. To be effective in such a world, we are told, requires us to be centred in ourselves, with a dominating, separative note enabling us to hold our own power and impose ourselves on circumstances and others.

Transcending separateness

I believe that the principles of the person-centred approach can enable people to move through negative conditions of worth and false conditioning. My concern lies in whether false conditioning is recognised as a possible block to growth and distinguished from negative conditioning. I feel sure that the actualising tendency holds the promise of taking the individual through the barriers of false conditioning, through the stage of 'separative humanness', towards that which transcends it. If we have within us a spiritual core, and an awareness and expression of that spiritual core is a key component of our wholeness and completeness as persons, then surely the actualising tendency must move us in that direction. We need to be open-minded, I think, as to just where the actualising tendency can take us. We should certainly beware of setting limits through our own mind-sets that are the product of false-conditioning to an identity with a separative sense of self.

Some argue that the idea of person-centredness is only applicable in the therapeutic encounter and not in the 'real' world, that the two can be, maybe should be, kept separate, with the therapist applying person-centredness to an encounter in order to stimulate constructive personality change. In response to this, first of all I do not believe you can throw a psychological switch to turn person-centredness on and off. Either the relational values and principles of the approach are internalised into your being as a functioning person, or they are not. And if they are not, then to portray them in the therapeutic encounter as if they are is at best disingenuine and at worst a deception. It is certainly incongruent and would therefore not fulfil the requirement to provide congruence in the therapeutic relationship.

There is also something alarming about seeing the principles of the person-centred approach as being something that you *only* apply in therapy in order to help someone. Yes, you may do this, but in so doing you deny the wider possibility and scope of the relational principles that are the foundation of the approach. I believe that this perspective is rooted in the very false conditioning that denies fundamental connectedness. To be person-centred outside of the therapy room would be to directly challenge the values that are dominant in much of the world. Consumer capitalism, dictatorship, all forms of totalitarianism, even representative democracy (a more person-centred approach would be a 'participative' democratic process) would all have to be challenged.

To bring the relational principles of the person-centred approach into the world at large would mean a fundamental shift in emphasis in the values that largely direct human behaviour and activity. And the interesting thing is that as more people realise the impact of their actions on others, as more people feel the need to change their lifestyle to reflect the growing sense of responsibility they are experiencing towards others and to the planetary eco-system, this evolutionary step is actually starting to happen. It may not be called 'person-centred', it does not need to be. But once you begin to question your actions and attitudes because of the negative or damaging, or growth-constricting effect it has on others, then you are part of the change. You are moving away from the 'sense of separateness', transcending this, if you like, by a greater sense of connectedness with, and responsibility towards, others. And today that means in a global sense.

The awareness we need to function in our increasingly global world requires us to expand our sense of personhood, and for this I believe there is a requirement to enter and more fully engage with a sense of transperson-centredness.

13

Person-centred Psychology: ... to the Transpersonal

Soul-centred Psychology

I would like, at this point, to introduce a perspective that has been described as *Soul-centred psychology* and links to the many references I have made within this book to the idea of the Soul. The Soul is to be regarded as that deeper, spiritual aspect of our nature that links us to spiritual realms and to God's Plan and Purpose for humanity. It acknowledges our spirituality, and the idea that we have, at the core of our psychological make-up, a Soul, a spiritual consciousness that transcends our normal awareness yet which we can access, experience and draw on. It can be thought of as embodying qualities of love and compassion, existing in a state of wholeness that in some way transcends the individualistic nature. Human beings are seen as part of an evolutionary process whose main thrust is the development and expansion of consciousness. The life process is about opening our hearts and minds to our Soul nature and allowing its essential qualities to find expression in our lives.

What force drives the growth and expansion of consciousness? *Love.* A quality of love that is rooted not in personal likes, desires and attractions but which is more of a universal force that bonds, unites and coheres. The Soul-centred perspective suggests that we are all subject to an evolutionary impulse that is driving us back to God. Each of us has within us God-like potential, a certain God-like immanence providing us with a path of approach towards the more commonly acknowledged transcending God.

I equate this God-like aspect with the Soul, the nature of which is surely love. This is by no means an idea special to any particular school of thought; it lies at the core of spiritual traditions around the world—that there is something deeper, some greater capacity and potential hidden from normal awareness For me, it is this crucial aspect of ourselves that holds the key to the basic trustworthiness of the human organism that Rogers emphasises. I feel uncomfortable with the word 'organism' which seems too biological, carrying with it an implication that we are nothing more than a physical body. I sense that Rogers, as his life progressed, realised that the organism was more, much more, than a biological entity. The passing over of his wife, Helen, had a deep effect on him and the following passage highlights his openness to larger possibilities.

> [my experiences] made me much more open to the possibility of the continuation of the individual human spirit, something I had never before believed possible. These experiences have left me very much interested in all types of paranormal phenomena. They have quite changed my understanding of the process of dying. I now consider it possible that each of us is a continuing spiritual essence lasting over time, and occasionally incarnated in a human body.[1]

Coming from someone who had previously sought to pioneer a strictly humanistic approach to psychotherapy, this is a challenging statement not only to those identifying their practice with the person-centred approach, but indeed to all within the humanistic and most other psychological traditions.

Many have experienced a yearning for a deeper connection and for more meaningful relationships in which this inner, spiritual nature can be present. This union or 'yoga' has been described using a variety of images, such as: the spark returning to the flame, the water droplet returning to the ocean, or the prodigal son returning to the Father's home. In the therapeutic relationship we can create an environment that touches this longing, creating a kind of re-membering of that which we know within us yet which we have forgotten. Something very precious can be created, that reflects a quality of being and wholeness that transcends the normal round of separative and selfish reactions in daily life. Therapy is not merely a technique for overcoming blocks and problems but offers an opportunity for growth beyond becoming a person. It offers the opportunity of Soul-centred living.

In reality we each have a good deal of experience that obstructs this kind of deeper contact. Blocks have to be faced and worked through in a way that is appropriate for each individual. There is a problem. The higher, spiritual component of the therapist has to be present in the therapeutic alliance if the client is to find stimulation and encouragement to reach into this area of his or her own being. When this is not present I would tentatively suggest that person-centredness can have a tendency to encourage self-centredness and to actually provide an

1. Rogers, C (1980) A Way of Being. Houghton Mifflin Co., Boston. pp. 91-2

obstruction to further growth. I have witnessed how the ideal of getting in touch with our own personal feelings can lead people to put these first, as if they have some paramount importance over the feelings of others. Is this not selfishness, sanctioned by therapy?

Spiritual teachings have emphasised the importance of right human relations as a fundamental factor in creating peace and establishing a new world order based on sound spiritual principles. Are not right human relations those characterised by genuineness, by unconditional positive regard (unconditional goodwill) and by an empathic understanding of each other? They can also offer us a view on how we should actually relate to others that we are seeking to help. In what are described as Rules for Harmlessness we read:

> "Rule 1. Enter thy brother's heart and see his woe. Then speak. Let the words spoken convey to him the potent force he needs to loose his chains. Yet loose them not thyself. Thine is the work to speak with understanding. The force receiving by him will aid him in his work.
> Rule 11. Enter thy brother's mind and read his thoughts, but only when thy thoughts are pure. Then think. Let the thoughts thus created enter thy brother's mind and blend with his. Yet keep detached thyself, for none have the right to sway a brother's mind. The only right there is, will make him say: "He loves. He standeth by. He knows, He thinks with me and I am strong to do the right." Learn thus to speak. Learn thus to think.
> Rule 111. Blend with thy brother's soul and know him as he is. Only upon the plane of soul can this be done. Elsewhere the blending feeds the fuel of his lower life. Then focus on the plan. Thus will he see the part that he and you and all men play. Thus will he enter into life and know the work accomplished."[2]

To be able to enter into a growth-enhancing relationship with another person it is clear that we must know ourselves and have purified our thought life. We must not seek to solve the other person's problems, but be a companion with them. We must be able to convey to them the love that we have for them in such a way that they know and experience it, and are strengthened thereby. Obviously the above goes beyond what is recognised by most forms of therapy. Yet it seems to me that therapy is a relational art that holds the seeds of becoming what the above rules indicate, given sufficient insight and awareness in the therapist. Soul-oriented therapy is a goal that will be reached and increasingly therapists and cli-

---

2. Bailey A (1934) A Treatise on White Magic. Lucis Press Ltd, London and New York. p.320

ents are themselves seeking this deeper reality. The strictly personal relationships are not fully satisfying and moments occur within the relationship that move beyond this experiencing. Clients are seeking this, therapists are being forced to develop in order to meet the demand. There is a process emerging which is drawing the Soul into the therapy room. The past has seen the main focus on the strictly personal (often emotional) level. Resolving issues on this level take the person along their journey, but it is still only part-way. The urge for something more deeply satisfying, whole and universally connected is a very real phenomenon.

If the therapist can touch the deeper soul levels within the therapeutic relationship then movement, growth, whatever we wish to call it can, and does, occur. To quote Rogers again:

> "When I am at my best, as a group facilitator or as a therapist, I discover another characteristic. I find that when I am closest to my inner, intuitive self, when I am somehow in touch with the unknown in me, when perhaps I am in a slightly altered state of consciousness, then whatever I do seems to be full of healing. Then, simply my presence is releasing and helpful to the other. There is nothing I can do to force this experience, but when I can relax and be close to the transcendent core of me, then I may behave in strange and impulsive ways in the relationship, ways which I cannot justify rationally, which have nothing to do with my thought processes. But these strange behaviours turn out to be right, in some odd way: it seems that my inner spirit has reached out and touched the inner spirit of the other. Our relationship transcends itself and becomes a part of something larger. Profound growth and healing and energy are present."[3]

It seems that from within his own frame of reference, experience and terminology, Rogers was indicating that he himself was entering into relationships having a flavour of those indicated in the Rules for Harmlessness in the sense of his entering into something larger and transcending.

The therapist seeks to be fearless and harmless. Both qualities are necessary. Fearless in terms of (the therapist) not screening out what may be said out of fear of how the client might react, and harmless in terms of being clear on their (the therapist's) intention to act in ways that enhance growth and not old patterns of abuse. Fearlessness links to congruence in the therapist, to the ability to be genuinely and fully present, even transparent, within the therapeutic relationship. Harmlessness concerns the sensitivity in which the therapist enters the client's

---

3. Rogers, C (1980) A Way of Being. Houghton Mifflin Co., Boston. p. 129

private world of experience, providing a presence, a witness that is there to support the growth process and not to stifle it where it gets painful, as it often does.

Let us return to the necessary and sufficient condition for therapeutic change as described in the previous chapter, but this time from the angle of the relationship between the soul and the everyday personality of the individual. Let us consider the Soul to be therapist and the personality as client:

1. Soul and personality are in psychological contact
2. The personality is in a state of incongruence, being vulnerable or anxious
3. The Soul is congruent or integrated in the relationship
4. The Soul experiences unconditional positive regard for the personality
5. The Soul experiences an empathic understanding of the personality's internal frame of reference and endeavours to communicate this experience to the personality
6. The communication to the personality of the Soul's empathic understanding and unconditional positive regard is to a minimal degree achieved

Is not the relationship we seek to build between our deeper spiritual Self (the Soul) and our personality built on the same necessary and sufficient conditions? Can we not think of the Soul-personality relationship as essentially a therapeutic relationship? Is not life, by it's very nature, taking us on a therapeutic journey that begins with personal growth, moves into spiritual growth and ends ... God only knows where? It is fascinating to reflect on the Soul-personality relationship in terms of being a kind of internal 'therapeutic alliance'.

Spirituality

Is there a higher level of humanness that transcends the therapeutic encounter and is unlimited by the false concept of separation? Have human beings the capacity to move into a different mode of engaging with the world? Within the therapeutic relationship it is the responsibility of the therapist to maintain a *way of being* with the client that can resonate with the client's higher potential. If this is lost sight of, the client could become locked into a self-centred stage of person-centredness.

I do not think this is good enough or, indeed, what Rogers would wish. I believe that he himself lived this journey and it was only in later years that he, too, began to break through into a more transcending realm of human interaction. Much of his work was modelled on his earlier experiences, and many of the papers that others have presented are based on Rogers' earlier ideas. I feel it represents only part of the journey. It needs to evolve further. Writing about the 'community-forming process', Rogers referred to characteristics of "transcendence, or spirituality". He goes on to say, "these are words that, in earlier years, I would never have used". His language moved as a result of his experiences – no doubt, 'moments of movement' – and clearly Rogers towards the end of his life was not the same man he was 50 years before. We need to be empathic to his personal journey and the meanings and values that emerged for him.

Without a sense of spirituality, person-centredness could become a partial therapy, leading clients into themselves, yet not acknowledging the possibility of enabling them to approach their Self, their true spiritual nature that is rooted in connection and love. The negative self-concepts are worked on, and resolved, but with what are they replaced? Maybe the result is a fully functioning, *biological* organism? Whilst I acknowledge that is an achievement in itself, I suggest we merely create a mechanism, a personality, through which something far more profound is waiting to find expression.

Can the client, when they experience acceptance and warmth towards their new, self-centred, self-affirming self concept, confront their achievement as a possible false conditioning and so journey on to discover something deeper? The spiritual aspect has not received enough attention within the person-centred world and it needs addressing if PCA is to remain a vital force in the human potential movement. If to be person-centred is to be open to experience and possibilities then should there not be more openness to a deeper connectedness, that is more rooted in our spiritual or deeper natures? Otherwise, we run the risk of becoming no more than fully-functioning biological organisms. The fully functioning person is surely more than this?

Rogers wrote in 1961 that as the individual "moves towards being open to all his experience" he "will be more balanced and realistic, behaviour which is appropriate to the survival and enhancement of a highly social animal"[4]. This sounds extremely biological to me. But by 1980 he was writing:

> Our experiences in therapy and in groups, it is clear, involves the transcendent, the indescribable, the spiritual. I am compelled to believe that I, like

4. Kirschenbaum, H and Henderson, VL (Eds) (1990) The Carl Rogers Reader,. Constable and Co., London. p. 419

> many others, have underestimated the importance of this mystical, spiritual dimension.[5]

There must have been a movement, an expansion to embrace other possibilities during those 20 years. Clearly he now recognised the need to explore the spiritual component of human experience. He was not alone among pioneering psychologists who pointed the way forward as being in this direction. Jung suggested that:

> We moderns are faced with the necessity of rediscovering the life of the spirit; we must experience it anew for ourselves. It is the only way in which we can break the spell that binds us to the cycle of biological events.[6]

Roberto Assagioli, founder of Psychosynthesis, wrote:

> [Human beings] can have the intuitive realisation of [their] essential identity with the supreme Reality. In the East it has been expressed as the identity between the Atman and the Brahman. In the West some mystics have boldly proclaimed the identity between [humankind] and God. Others have emphasised that life is One, that there is only One Life. But this does not mean that man's *mind* can grasp the wonder and mysteries of the cosmic manifestation. Only through a series of expansions of consciousness, only by reaching ever higher states of awareness, may he gradually experience *some* of these wondrous mysteries.
> Of such transpersonal possibilities the most enlightened men and women of all ages have given testimony, expressing them in basically the same way, above the differences and colourings due to individual and cultural conditionings.[7]

I would therefore like to suggest the need for a change in mind-set, a need to build upon what we already have by acknowledging that beyond person-centredness is something that might be called 'transperson-centredness'. It is a rather awkward description with too many syllables, yet I think it captures where the vision needs to be more strongly directed. It requires us to open our hearts and minds to greater possibilities and potentialities, to face a challenge that has been

5. Rogers, C (1980) A Way of Being. Houghton Mifflin Co., Boston. p. 130
6. Jung, C, (1933) Modern Man in Search of a Soul. Routledge & Kegan Paul Ltd., London. p. 122
7. Assagioli, R (1974) The Act of Will. Turnstone Press, Wellingborough. pp. 125-6

succinctly described by David Spangler as one in which people need to discover ways

> ... to utilise the energies of [their] life, of [their] consciousness, in such a way that [their] divine identity is revealed and [their] personality identity becomes illumined within that revelation and ceases to be a disunified element operating on its own.[8]

The core conditions—the three in one

I would like to return to the core conditions which, for me, each reflects a different quality of love and to suggest that they resonate with the triune nature of spirit. To be real with someone (congruent) is to express a loving *will*, a will to be true to ourselves and to that person; to feel unconditional positive regard towards someone is to express a loving heart, a love that acknowledges another person's own capacity for love at the core of their being; to strive for a truly empathic understanding of another person is an expression of a loving *mind*, stemming from an intimate knowing of the client's world.

Within some traditions, God is said to be triune in expression whilst singular in existence: spiritual will, spiritual love and spiritual mind; Father, Son and Holy Ghost; Shiva, Vishnu, Brahma. We have here a fundamental triplicity of creation. They form a triangle, a most ancient and significant symbol. Spiritual tradition speaks of the three in one, that the three can by synthesised into a unity that includes the three yet which is more; in effect it is greater than the sum of its parts. This unity is perhaps some greater quality of Love operating out of a deeper dimension. Can we find a parallel with the core conditions? I would suggest that we can, that the three core conditions are, in some mysterious way, contained in the one which Rogers calls *quality of presence*.

Such presence proceeds from a sense of unity between client and counsellor, from a condition of love that transcends personal desire and reaction. It is borne out of a wholeness: separation ceases to operate; negative and false conditions of worth cease to exist in the moment. A glimpse of a fuller Reality is gained that perhaps stirs hidden memories or sensitivities. Neither client nor therapist can ever be quite the same again. A moment of real movement takes place.

Rogers wrote of how when he was closest to his inner, intuitive self, to the unknown in himself, and perhaps "in a slightly altered state of consciousness"

8. Spangler, D (1978) Relationship and Identity, Findhorn Press, Forres. pp. 50-1

then whatever he did seemed "to be full of healing". I here repeat the quote given previously:

> Then, simply my presence is releasing and helpful to the other. There is nothing I can do to force this experience, but when I can relax and be close to the transcendental core of me, then I may behave in strange and impulsive ways in the relationship, ways which I cannot justify rationally, which have nothing to do with my thought processes. But these strange behaviours turn out to be right, in some odd way: it seems that my inner spirit has reached out and touched the inner spirit of the other. Our relationship transcends itself and becomes a part of something larger. Profound growth and healing and energy are present.[9]

This does not mean that a poor self-concept disappears for good, but its disappearance in the moment offers an opportunity to experience a deeper, less self-centred, transpersonal way of being. How might we describe such experiences? They are beyond words: only silence seems to be appropriate. Such moments are rooted in a quality of Knowing, Loving and Being that is more God-like than human. It is important to emphasise that such experiences, if they are to have relevance to human evolution, must work out through the person. This has been highlighted by one of the forerunners in the use of the term, 'transpersonal' – Dane Rudhyar:

> I have spoken of the "transpersonal way"; but in using the term, transpersonal, I am referring to an activity that is focused through the person, not merely to a reaching out beyond or above the personality.[10]

In conclusion

We are struggling within a world of conditioned experience and identity to discover who or what we truly are, to break free of the self-concepts, the negative and false conditioning that bind us to a world of separateness. It is a world that, according to Rogers, is pervaded by a formative tendency and our goal is to allow ourselves to participate fully and consciously in the formative process. Can we make constructive use of the actualising tendency that Rogers described? Can we

9. Rogers, C (1980) A Way of Being. Houghton Mifflin Co., Boston. p. 129
10. Dane Rudhyar, Directions for New Life, quoted in Twelve Seats at the Round Table, by Edward Matchett and Sir George Trevelyan (1976) Neville Spearman, Jersey. p. 15

find ways to co-operate with this all-encompassing evolutionary impulse, the formative tendency – the urge towards fuller functionality and growth/expansion in consciousness? Can we tread a path through life that will lead us from darkness to light; from the unreal world of negative and false conditioning to the Real world that in some way proceeds out of our spiritual core; from the death of conditioned existence to the immortality of Soul-centred personhood?

> A new evolutionary level of human consciousness is even now seeking to find in men, [women] and groups everywhere focal points for manifestation; and at that level their keynote is love—agape, the love of the companions through whom Man may find agents for synthesis and for the harmonisation of his myriads of potentialities through co-operative action ... *now* is the time for mutation. It is a basic mutation, just as fundamental as that which led tribal man to form civilisations, where minds and individuals learned to think, to question, to gain personal independence, to assert their ego-will and yearn for personal power and lustful excitement.[11]

We are left to ponder what it is that we essentially are. In our present time, more and more people seem to be searching for greater meaning in their lives, having found their life-experience barren or simply too painful to bear. Why do so many people today turn to therapy, join self-awareness groups, meditate or generally involve themselves in the human potential movement?

There is a growing belief in a reality that in some way transcends and, by its very nature, gives meaning to our experience of life in this world. What is going on? What is it they seek? And what happens when they find *it* and know *it*, and realise that *it* is simply and truly themselves, rooted in a common core of divinity that had been lost sight of? What happens when the divine is actualised in people? Could it be that they will experience fresh motivation to bring qualities of love and compassion into the world? Might they become more responsive and responsible individuals towards human and world need? I would suggest that somewhere in this experience comes a realisation of what it is to be truly *transperson-centred.*

Many training courses in therapy acknowledge the fundamentals of what Carl Rogers was suggesting, yet few recognise the spiritual or transpersonal depth that is present when you look for it. He himself was seeking strictly human answers to questions and his theories developed out of the humanistic tradition of psychol-

---

11. Dane Rudhyar, Directions for New Life, quoted in Twelve Seats at the Round Table, by Edward Matchett and Sir George Trevelyan (1976) Neville Spearman, Jersey. pp. 15-16

ogy. Most of his life was not spent advocating a world-view involving a spiritual component. Yet as his life proceeded he glimpsed that something other, that something deeper, that sense of transcendence that the mystic or meditator works towards through spiritual training.

> "Our experiences in therapy and in groups, it is clear, involves the transcendent, the indescribable, the spiritual. I am compelled to believe that I, like many others, have underestimated the importance of this mystical, spiritual dimension.[12]

Therapy is on the threshold of moving more forcibly into relationship with the spiritual dimension. More clients are wanting to make this journey and are seeking therapists to be companions on that journey. People are finding themselves growing through the experience of 'right relationship' in the therapeutic setting. It was the late Ian Gordon Brown who, having established training in transpersonal psychology, commented to the effect that there are no new pioneering thinkers in psychology today, the real pioneers are the clients who are breaking into new realms of being.

We can all grow into new perspectives, into new ways of being in our process of becoming more than what we currently experience ourselves to be. As we view the jigsaw of life there is, I believe, a vital need to include within it the realm of what I am calling the transpersonal, bringing into focus the spiritual dimension and potential that, it seems to me, we are only starting to explore.

---

12. Rogers, C (1980) A Way of Being. Houghton Mifflin Co., Boston. p. 130

14

The Jigsaw of Self[1]

It is my belief that we, as human beings, have greater potential to resolve our own difficulties and problems than we are often credited with. I believe that life-experience can undermine our sense of self, leaving us weak and unconvinced of our own abilities. The image of the jigsaw already described in earlier chapters can also be applied to the individual so that we can consider the image of a jigsaw as representing the person, the sense of self that we carry with us into each day.

I am convinced that we begin our lives with many potentials and talents. These may (or may not) develop and find expression as we move through life's experiences. Their emergence can be hindered as we develop a view of ourselves based largely around the opinions and feedback we receive from significant others, some of which generates a negative self-concept or poor self-image. We may find ourselves conditioned into ways of being, into ways of behaving and of seeing ourselves that are more concerned with fulfilling other peoples' expectations than discovering our own worth and genuine self-expression. We can think of this as being represented by a jigsaw whose pieces have been put together in such a way that the overall picture cannot be seen, creating a disjointed image that does not measure up to our full worth as a person. We are often caught up in putting much of our energy into creating someone else's picture rather than our own.

If we think of ourselves (albeit rather simplistically) as a jigsaw put together over the course of our lives, many of us will feel, I am sure, that we have got some of the bits in the wrong places. If the jigsaw represents our sense of self then we may continue putting down pieces in places that, whilst producing an image of colours and shapes, obscures our potential as a fully functioning person. We cannot be the person we have the potential to become if the pieces do not fit together

1. This chapter in a slightly modified form was published in Bryant-Jefferies, R (2003) *Counselling a Recovering Drug User*. Radcliffe Publishing, Abingdon.

in a way that makes us feel whole and fully present. We may be left with a feeling of inner fragmentation.

We may go through life accepting our sense of self, the image that we have created metaphorically in the jigsaw, subtly conditioned into believing it to be a truthful representation of who we are, or rather, of who we have become. And in a real sense, it is truthful. We can feel congruent to that image, seeing it as a mirrored reflection of how we see and experience ourselves, and confirmed to us through the feedback that we receive from those around us. Our 'locus of evaluation', as Carl Rogers termed it, has become centred outside of ourselves, invested in the opinion of others. We may continue to function well in this condition, feeling comfortable with the sense of self that we have created, even if it is obscuring our ability to engage with a state of fuller functionality as a person.

Yet we also know that many of us enter into a phase, or phases, of life when we question our sense of self; times when we begin to wonder whether we truly are being ourselves or living out the hopes, expectations, dreams and fears of the significant others that make such a deep impression upon us. It may be that the person that we have identified ourselves as being for so long, the self-concept that we have created for ourselves, is somehow not working any more; we may be breaking down under the pressure that life's demands have placed upon us. Or we may feel an overwhelming sense of being constrained by patterns and habits of living and being that we no longer find as satisfying as they once were. We can feel out of step with ourselves, letting the old habits and patterns live on (keeping to the same image of the jigsaw) yet feeling somehow it is not us, that there is something out of place. We are beginning to feel that the image on the jigsaw of self is not as we wish to see ourselves, or feel we want to be. Another way we might experience this is through the sense that life is taking us in a direction that we do not feel is ours to take.

It is at this point that a person may find him or herself facing a crisis of decision:

- whether to shore up the self-concept that has developed over the years and with which they are now so familiar, even though it no longer feels right, in the hope that it will all be OK and that it is just a 'silly phase' they are passing through;

- whether to face the possibility that through conditioning they have lost touch with the potential as a person that they might grow towards becoming.

This may seem a stark and simplistic choice to be made between two extremes. In reality, I believe, they do not exist as extremes in isolation but rather as extreme ends of a continuum. The reality is often that we may wish to accept certain aspects of who and how we have become whilst questioning other areas of our natures and habits of living. It is often parts of our nature or behaviour that we feel unsure or uncomfortable about although it can be the situation that we are questioning our whole sense of self. We may choose to seek help, guidance or direction and this may involve a decision to enter into a therapeutic relationship with another person in which to make sense of and grow through the crisis. It is here that I wish to return to the image of the jigsaw.

Client-centred Therapy

There are many approaches to counselling and psychotherapy. I wish to focus on client-centred therapy which I have referred to in previous chapters, but to recap it is an approach in which the therapist seeks to communicate empathy, congruence and unconditional positive regard towards the client. The client is offered freedom to explore whatever they wish to focus on without the therapist seeking to make interpretations or judgements as to what choices should, or should not, be made. A core attitude within the therapist is a recognition that the person they are with can be trusted to experience growth where these core relational qualities (sometimes referred to as 'core conditions') are present. Yet it has to be more than a recognition. The trustworthiness of the individual and the presence of these core conditions have to be communicated genuinely by the therapist and received accurately by the client. It is not the case with this way of working that the therapist seeks to "sort out" or "heal" or "make better", but rather the intention is to offer a genuine, sensitive and open relationship. It is from this accepting experience that the client may then undergo a psychological process moving them towards healing or resolution of difficulties. In theoretical terminology, it provides the client with a relational experience such that constructive personality change may occur. It means the person becomes more authentically who they are and more able to function to a fuller degree as a person with all the qualities and attributes that they have at their disposal.

Imagine yourself as the client. You have come along because of difficulties in your life. You are not coping well and you want to make sense of what is going on and to find a solution. In terms of the jigsaw metaphor, the pieces have not been put together as they might have been, they have not been assembled in a way that might have enabled you to feel more whole and genuinely yourself as a fulfilled person. Yes, some of the pieces represent aspects of the you that you want to be, whilst others have been formed and introduced by other people, wanting you to grow and develop in their image, and not in your own, and in ways to meet their needs, not yours.

The therapeutic process begins and you talk about yourself; how you see yourself; the situations you feel you handle well, and those you do not; the highs and lows in your life; your hopes and fears. By talking and exploring these facets of your life and experience you 'externalise' them, you externalise a picture of yourself from which you can begin to see and experience the whole picture that you have created more clearly. You are in a position then to question how some of the pieces have been fitted together. This is not analysis at a distance. Feelings and thoughts are engaged with, and are often lived in the therapy session. What seems to be important is that they are "put out" and the therapeutic relationship is such that they can be explored, understood and reintegrated – put back – yet with perhaps an added or different meaning as a result. And some may be rejected, seen as a direct product of past conditioning from others, to be now discarded as not conveying the personhood that is now emerging.

In the therapeutic relationship you find your sense of self changing as you feel heard and listened to, valued and accepted as a person in your own right by someone who is endeavouring to be real and fully present in relationship with you. Pieces of the jigsaw are seen more clearly as being in the wrong place, symbolising those elements within your own nature and self-concept that are the effect of conditioning in your life and which have had a bearing on your self-concept. The safety and trustworthiness of the relationship enables you to explore more freely, experimenting with new ways of being that seem to be more you, releasing emotional and intellectual blocks to change.

The role of the client-centred therapist is not to point out or to suggest which bits to move in the metaphorical jigsaw. This is crucial. The process is concerned with enabling you to trust your own judgement and insight, to believe in the validity of your perceptions and interpretations. You, the client, are trusted to see for yourself. The time to question the placing of a piece within your personal, metaphorical jigsaw of self is when you, the client, acknowledge that there is something about that piece that it is not right. The therapist will help the exploration of what a piece may mean to you, why it has been placed where it has, why it is no longer seen to fit in with your changing sense of self. The therapist will not, however, tell the client to change it. The piece can only be changed when the client knows within him or herself that it is no longer right, that the facet of themselves that it represents is no longer a sustainable part of his or her self-image. At this point they may have begun to develop a sense of what it is to be replaced with, but not necessarily so. Sometimes we simply know what does not feel right before we create the space within our psychological functioning to then engage with what actually does feel right.

Limitation and Expansion

I do not want to encourage, however, a view that is too rigid of the process of solving or resolving—I am not sure which is the most appropriate word—the jigsaw of self. I don't want it to sound mechanistic. It isn't. I am not even sure whether it is a linear piece-by-piece process; maybe we get breakthroughs and pieces come together in groups, clumping together to generate a new self-image in a particular area of our life. Perhaps there are sudden shifts as we see ourselves in a fresh light and reinterpret our own experiences from the past and the present. Perhaps these significant changes might be seen to correspond with Carl Rogers' idea of 'moments of movement', times when something shifts within the individual's consciousness. Using the jigsaw metaphor we might suggest these to be times when a piece of ourselves falls into place and there is a satisfying sense of 'fit'.

It may even be that in some mysterious way parts of our nature are assembled before they are positioned within our natural way of being, represented by how, when working on a jigsaw, we might group pieces to complete a small section of the picture before introducing it into the larger picture itself. Can this process occur psychologically? Perhaps it can where we act 'as if' we had a particular quality, testing out an element of our potential prior to that final acceptance of "yes, that really is me".

Of course, the jigsaw image has its limitations. It may be that it should be thought about in other ways to become a more realistic reflection of the self. I am unsure, for instance, whether the pieces should have fixed shapes, patterns or colours, or whether they may themselves evolve. I also want to know whether there is a limit to the possible number of pieces available, whether we ourselves have infinite possibilities or whether there is a fixed limit to who we can become. I wonder, too, whether the pieces that are considered to be in the wrong places need to be placed elsewhere, or whether they need to be discarded, no longer part of the overall picture. And what does it mean, psychologically, to discard a piece of the jigsaw of self? Or is it not so much a case of discarding pieces as allowing or enabling pieces to evolve within us?

I also find myself wondering whether the image itself contains the seeds of new images that will emerge in time, a kind of fractal reality which, when the image is completed, enables the individual to move deeper into themselves or maybe gain a more expansive sense of self. What does 'expansive' mean in this context? Does it indicate greater congruence in the person as they are, or is there an added sense of the potential to become more than their current, congruent self?

Could it be that from an early age we contain within us the seeds of our becoming, yet the journey we take and the self we create as we move through life ensures that the final outcome cannot be known until it is reached? Is it that in life there is not so much a goal, but more of a direction? Carl Rogers wrote of an 'actualising tendency', an urge to grow towards fuller functionality as a person. Perhaps our experience of the journey through life is more significant than being overly concerned about reaching the goal? I can only wonder as to the undiscovered potentialities of what it is to be a truly fully functioning human being. And it seems likely that such full functioning must include greater relational capacity and sensitivity to the spiritual It seems to me that we may make more wonderful and unexpected discoveries through the exploration of inner space and how this interacts with and through external relationship(s) than we will in our exploration of outer space! Or perhaps the two must go hand-in-hand. We will only fully grasp cosmology and the wonders of the universe – from star systems to quarks – as we cultivate new ways of thinking and relating to the world both around and within us.

Towards completion

It may well be a slow process to rebuild our jigsaw of self, or it may prove to be quite rapid. Some people may only need to change a few pieces, others may require a great deal of change, many pieces needing to be moved or removed. It is my belief that by engaging in a client-centred therapeutic relationship the jigsaw of self can change and evolve into a clearer and less distorted image of the person. It may not be possible to get all the pieces in the 'right' places through one series of therapy sessions. Breaks may be needed to assimilate and consolidate the new arrangement, to put it to the test, if you like. It may be that particular areas of the picture are concentrated on as these are areas in which the client is aware of a need to change, or at least to revisit and engage with. It is likely that even by the end of life the jigsaw may not be fully completed. Indeed, we might consider ourselves extremely fortunate if it is! Yet it is my belief that the more we can solve the jigsaw of self, ensuring that more and more of the pieces are appropriately positioned in relation to each other, the greater will be our capacity to live enriching and nourishing lives, and be more of the person that we have the potential to become.

And, of course, we do not just build our jigsaws of self in the therapy room. We are doing it all the time, everywhere we go, with every experience that we encounter. And what is perhaps significant in our time is that for more and more people their jigsaw of self is not only taking place within and without the therapy room, but it is taking place in a global context. This is new. This is different. This, more than anything, is what makes our time in human history particularly significant. To return to the jigsaw metaphor, we are now in a position to more consciously shape our pieces with more of an awareness of the many other pieces and the groups of pieces that are forming than ever before. The evolution and development of the part becomes informed by the increasing presence of the emerging whole.

◆ ◆ ◆

You look down and you see the surface of that globe that you've lived on all this time, and you know all those people down there, and they are like you, they are you, and somehow you represent them. You are up here as the sensing element, than point out on the end, and that's a humbling feeling. It's a feeling that says you have a responsibility. It's not for yourself. The eye that doesn't see doesn't do justice to the body. That's why it's there; that's why you are out there. And somehow you recognize that you're a piece of this total life. And you're out there on that forefront and you have to bring it back somehow. And that becomes a rather special responsibility, and it tells you something about your relationship with this thing called life. So that's a change. That's something new. And when you come back there's a difference in that world now. There's a difference in that relationship between you and that planet and you and all those other forms of life on that planet, because you've had that kind of experience. It's a difference and it's so precious.[2]

2. Russell Schweickart, US astronaut. Quoted in *The Home Planet.* KW Kelley (Ed) Addison Wesley Publishing Company. (1998)

All is One

One is All and All is One:
The Man, the tree, the earth, the sun.
Forms may differ in shape and tone
But none of these can stand alone.

Connections span both space and time.
History's bell is heard to chime.
Actions past are never lost;
They touch us now, we count the cost
Of moments in deluded state
When jealousy, envy, greed and hate
Took firmer grip upon our souls.
We lost the truth, we lost the Whole.

And yet not lost, just veiled from sight,
The truth awaits inclusive light.
Connections are, though we deny
Reality on which they lie.
And as the veils begin to clear
The threads of Life once more appear.

Yet each can stand apart or whole,
Reflected measures of the Soul.
The choice is ours, the need is great,
We can't afford to separate
And cling to veils that serve to blind
Our eyes to One Who stands behind.

15

The Thinning of the Veil

It is my belief that we are on the threshold of significant changes within humanity, and that the current evolutionary struggle is to do with acknowledging ourselves as a having a reality beyond the physical world of our sense. As part of this process more and more people are undergoing experiences that tell them, in a personal and immediate way, that their consciousness is not limited to a brain function. The huge numbers of people who testify to mystical, out-of-the-body and near-death experiences are an indicator of the way human evolution is moving.

It also seems to me that very often in life the external world reflects processes that are taking place on inner, subtler levels. Whilst reflecting on this I began to wonder at one of the key issues today: the thinning of the ozone layer. This protective shield has provided a barrier to excessive amounts of ultra-violet reaching the planet surface for millennia. Yet today scientific investigation has shown that it is dissipating in many areas at an alarming rate.

This break up of the ozone layer can be regarded as an 'unnatural' process in the sense that it has stemmed from humanity's so-called 'unnatural' activity of producing chemicals and other pollutants. It can be seen as a distortion of previously 'stable' atmospheric processes, imposed on the atmosphere by activities that have little or no regard for nature.

Yet we have to also acknowledge that it is nature that is creating these processes. Human beings, endowed with creative thought and ingenuity are a product of nature. The chemicals produced may not exist naturally outside of human activity, but as they are a product of us and we are part of nature then they must surely be regarded as being natural? Just because we do not like the effect does not mean we have to label it as unnatural. A lot of deadly substances exist naturally in nature.

By seeing the thinning of the ozone layer as an effect of unnatural processes can encourage us to look for and only see negative effects. In a world dominated by materialism, what could be considered more negative than the destruction of form? Research points to a rise in the incidence of skin cancer, which is already happening in some parts of the world. We have over recent years experienced an increased sense of burning from the suns rays. There is concern being voiced about the effect that the wider and more intense range of light reaching the surface of the planet will have on people's eyes.

There is also concern about what effect it might have upon vegetation. Will greater amounts of ultra-violet light affect the process of photosynthesis and, if so, will it prove damaging and what will be the long-term consequences? Will certain species die away? Could it have a dramatic influence on human beings through the food chain? There remains a lot of uncertainty and a great many questions to ask and to answer. What we can agree on is the simple fact that what will happen – is already happening – is that a quality of energy, of light, previously veiled from us to some degree, is now being released into our world in a new and powerful way.

The ozone layer, as a veil shielding the Earth from excessive radiation emanating from the sun, could be seen as a symbol of the inner veil that shields the developing individual consciousness from the full brilliance and potency of the light of the Soul or Spirit. Certainly the sun became fixed in the hearts and minds of many ancient traditions as a symbol of the Creator, as a focus for reverence and worship.

Could the thinning ozone layer be an external indicator of an inner process of change, a thinning of the veils between the Soul and its mechanism, the human personality and form? As we open the inner eye to the light of the Soul we may increasingly experience our outer eyes awakening to a new sensitivity to the widely acknowledged energy body – the etheric or vital body – that underlies the physical form. It is interesting to note, I think, that the process of thinning in the atmosphere is being fuelled by what is going on below, and that this parallels the process that take place in the individual. We disrupt and burn through the veils as energy patterns change due to shifts in consciousness. We burn the veils upwards towards the light in our selves to enable greater light to flow into our natures. It is a beautiful correspondence to the process that is taking place in the atmosphere.

Of course, this process involves danger. If an individual opens up to the intense inner light too soon, before the form can take the stimulation and absorb the inner radiation, damage can occur. There are exercises to force open channels to generate an intensified flow of energy and many have suffered as a result of burning through protective veils. Safety lies in ensuring that it is a gradual process. Anything else is to play with fire—literally.

Yet evolution is moving us on towards this breaking down of the inner veils. Many see this time as being one of tremendous opportunity for growth and for a real shift of energy state within humanity, sometimes referred in esoteric teachings as being an initiatory experience. Is the thinning of the ozone layer in the heavens a sign of this change? Can we apply the ancient axiom of 'As above, so below'? Could it be more than a sign and actually be part of the very process that is stimulating change?

What do I mean by this last remark? Human form is said to have developed through adaptation to the environment. Today, the environment in terms of the light reaching the Earth is changing. Could it be forcing us to adapt in some new way? Might it be stimulating the development of a sensitivity to a wider frequency of light? Is it possible that the thinning of the ozone layer and the resulting inflow of ultra-violet light is, in time, going to stimulate or induce etheric vision – the ability to see the energy fields underlying all created forms?

Light is a very strange phenomena. It has been the subject of so much debate and research. Recently I learned of a condition where people are blind when their eyes are exposed to sunlight yet they can see in artificial light. What is the dominant factor that governs the development and experience of sight? Is it the nature of the light, the sensitivity of the eye, or the ability of the brain to register and interpret the reality of the images it receives? The images we see of the world 'out there' are very limited. We have a very small window to look through on to the world. Could it be that the window is widening?

Consider Darwinian theory. The species survives through a process of adaptation and transformation that is stimulated by interactions with the environment and passed on and reinforced through a process of genetic inheritance. Those species that can best adapt and can pass on their adaptation survive. Are we one the verge of a major shift in human adaptation? Are we going to find in the decades ahead of us that more and more people are going to develop etheric vision, or have bodies which, through genetic inheritance, will contain eyes and the necessary brain functioning that will enable them to be sensitive to a wider frequency of light? Might this not also be part of a greater process of actually shifting our focus, individually and collectively, on to the etheric energy world? Could this then have the knock on effect of encouraging us to more fully co-operate with the energy processes of our planet?

To begin to think this way we have to look beyond the negative effects on form of these changes. It can, at this stage, only be a working hypothesis. We do not know if this is what is happening and what will be the effect. But suppose the hypothesis is right? Suppose changes will or are occurring in the process of human sight? If people naturally find themselves able to see energy fields, able to see the energy body that underlies and sustains our physical bodies, imagine what a breakthrough it will be. The world will change. Our individual worlds will change. Nothing will ever be quite the same again. We will then *know* that all is energy because we will *see* it. Separatism will be lost sight of by more and more people because it will simply no longer exist in their frame of reference, it will have no reality for them. We will know that we are connected at an energetic level and we will know that energy follow thought. We will see this process occurring before our very eyes.

Fantastic? Fanciful? Impossible to imagine? The testimony to etheric vision is already on the increase. Many children naturally see energies that adults are blind to. As etheric vision becomes a more accepted reality – should I say, *the* accepted reality, change will be enormous. Think about it for a moment. Think of the implication on human attitude and activity, on human thought and feeling. Think of the incredible implications of the human eye developing sensitivity to subtler energies and the human brain adapting in such a way that we experience vision of those energies.

What else might we see? What else is 'out there' that is currently invisible because it exists on subtler wavelengths? What other things might the brain register once sensitivity is developed? I mentioned in an earlier chapter the idea of the reappearance of the Christ impulse. Would such developing 'light-sensitivity' ensure that this impulse will be sensed and known by – to borrow an Christian phrase – "those who have eyes to see"?

Of course, there may be other forces at work that block this development. If there is an evolutionary process taking place within the eye and the brain, what of the substances that we put into our bodies. Could they block this development? We do not know. It may be that whilst the emphasis on the threat of chemical warfare is on the external world, it could be that the real chemical war is the one going on inside our heads.

There is also another angle on this as we see developments in computer technology. Might we find implants developed that might also enable people to 'see' beyond the current visible spectrum, not only towards the ultra-violet but also the infra-red? I am not sure if this is perhaps the best way, it does not seem quite such a 'natural' development somehow, but I am sure the potential is there. Again it will impact on human consciousness.

Getting back to the more natural process, of course all of this may be an idealistic and overly optimistic interpretation of the thinning of the ozone layer, factually or symbolically. We must each make up our own minds. Whether true or not, the hypothesis is, I believe, certainly worth pursuing and thinking about. Who knows, perhaps if energy does follow thought, then our positive thoughts towards the very idea of fresh, more, different – what is the best word to use? – light being released into our atmosphere may contribute in some mysterious way to encouraging the process? It might even be part of the process and may lead us to clearer insight into the jigsaw of life, into the nature of ourselves, the world and of the created universe. We may have much to learn about the effects of the presence of more light – in our inner and outer worlds – on human vision and consciousness.

16

Of Life and Death

And so to the really big question. What is Life and what is Death? How can these essential aspects of human experience relate to the notion of a jigsaw of life, or indeed, as we saw in the first chapter, a truly cosmic jigsaw? I think we have to ask ourselves some questions, and I do not propose to be able to give the answers. Indeed, sometimes we learn from holding a question open rather than closing it by a belief or an idea that may be right, but also may be wrong.

- Is there a purpose to life? And by this I mean is it all more than some chance happening? Yes, there is intelligence and complexity within creation, we cannot doubt that, but is there an actual purpose to it all? Are we heading somewhere? Is there indeed a final picture that we are assembling, or rather that we are each part of the assembly?

- If there is a purpose to human life, then what is it? Is it to develop human potential to its fullest? And if so, what would that look like? How much further have we to travel on this journey? Have we only just begun, or are we nearing completion, or are we somewhere midway?

- If we are on a purposeful journey, how can we contribute to ensuring that we continue in the right direction? What will enable us to achieve our fullest potential and, if we were to achieve this, what would the world be like? Think about it. Just what the world be like if human beings achieved their fullest potential or full-functionality as a species?

- If there is a 'fullest potential' to be achieved, what would it be? Would it involve us all having mega-brains of some kind able to process even more information than we do today? Or will we need to develop other qualities and capacities, and I am thinking here of the heart and of our capacity for compassion and for love; or of the human will aligning itself more with some greater Will and Purpose.

- And what of the spiritual realm? Does it exist? Or is it a figment of our imaginations, or experiential by-products of chemical and electro-magnetic interactions within our brains? Are we brain-bound beings, or is there more to who and what we are?

And, of course, there still remain two further unanswered questions:

- Is there just one life that we each have and that's it – we make our mark on the world for others to follow and hopefully build on our achievements? And we may continue to exist, as many believe, some within a Heaven or a Paradise, but from there can we influence human evolution, or are we separated from it?

- Is there a seemingly infinite number of souls (assuming that the soul continues to exist after physical death), or a fixed number that rotates through physical form? Do we come back ourselves in some mysterious way to carry on the work? If there isn't, as many also believe, just one life, and if we have therefore pre-existed in some mysterious way, then what impact would our past lives have on our present, and on the psychological processes that give us our sense of self?

This is a short chapter. I am not going to try and answer these questions for you. All I would ask is that you approach them with an open mind and make sense of them as you will, drawing on whatever scientific, religious or other ideas and guidance as you choose, along with your own experience and intelligent reasoning. What I think we should be aware of is that we may not know the whole Truth as yet. The Prophets have told us what God has chosen to reveal at certain times in history, but has He chosen to tell us everything? Science has made its discoveries and will surely discover more. Inspired individuals have glimpsed a greater reality and sought to convey that through many mediums: art, music, poetry and other writings. We all have much to reflect on as we seek to make sense of the Jigsaw of Life.

17

Conclusion

Let me draw together some of the threads from the chapters within this book, to highlight once more some of the qualities that I have identified as having what I believe to be some particular significance as we make our way towards the future. We have encountered a range of qualities and concepts within the preceeding chapters and here I am simply going to list them and leave you to ponder over what they mean for you and for humanity as individually and collectively we seek to find a way forward towards a better world and one that, perhaps, more truly reflects that final picture in the Jigsaw of Life:

- Spirituality
- Goodwill
- Responsibility
- Integrity
- Community
- Service
- Compassion
- Caring
- Right Human Relations
- Unanimity
- Prayer and Invocation
- Wisdom
- Trust
- Transpersonal
- Soul

- Interculturalism
- Equality
- Diversity

An important word is missing from this list—love. I wish to end this chapter with a quote from a passage that I wrote a few years back which I believe captures something of the challenge of finding our true place within the Jigsaw of Life. It is a passage entitled 'Call to Love'[1].

Call to Love

In the final analysis we have all got to find ways that enable us to know ourselves. We have to know who we are, why we function the way that we do and taking a firm grip on our often unruly and separative nature, drive it in the direction that we honestly and truly want to travel in …

On my journey of personal growth, if I have learned anything, it is this—the Soul, the spiritual Self, is my Reality. All else is a field of experience that can be transformed into a field of service through the presence of the Soul. Growth is for me about the struggle to rediscover my identity as the spiritual Self, so that I may then truly live to some greater purpose free of the limitation of a sense of separateness.

We can no longer afford to allow our creative expression to be dominated by petty personalities and greed. We have billions of people to feed, billions of people to house and to clothe, billions of people that, like us, would like to live with dignity.

[I believe in] the importance of having some form of personal contact with people in other countries. I remain convinced that we respond differently to news, particularly bad news, when we have a personal contact in a trouble-spot. I know how easy it is to experience what has glibly been called 'compassion fatigue', but can 'compassion fatigue' exist for us when we know someone living in a place where painful events are occurring? Our sense of humanity will not allow it.

---

1. Francis, L and Bryant-Jefferies, R (1998) The Sevenfold Circle: Self Awareness in Dance. Findhorn Press, Forres. pp. 110-112

Yes, it hurts to see some of the scenes that we do on our televisions, or to read reports and see pictures in our newspapers—and these are only the ones that we are allowed to see, the really sharp images never reach us. We need to stand in solidarity with those who suffer daily. This is my struggle as a counsellor and therapist, but I know that it helps.

I ask myself, "how can they feel my love if I cannot feel their pain?" This seems to really sum it up for me, somehow. Think about it. It is all very well thinking light and love and believing that this is going to change everything. This is good and positive to do. It contributes, but alone it is not enough. We need human-to-human, person-to-person connections if our hearts are truly to be effective in allowing a flow of love to occur. We have to engage with people, touch them and be touched by them. We must see their woe and then, in communion with them, let the light and love of the Soul blaze forth. Do not just imagine it, *live it, be it, make it real.* Spiritual knowledge always brings responsibility to live within our own circle of expression and relationship. The world is today that circle.

It is hard and it is painful. It is easy to feel dragged down by what we witness in our troubled world. Yet we cannot look the other way. We have to be hurt too. Nobody is an island, however much we might try to insulate ourselves. We are one human family. The implications of planetary connectedness are awesome. Energy follows thought. Energy follows feeling. The sense of separation only exists in our minds and feeling natures that prefer to be absorbed in our own comfort. "Comfort", I remember reading many years ago, "is the graveyard of the Soul".

Many believe that the Christ will reappear in some form, perhaps to offer another opportunity for humanity to hear the truth's that we need to hear. Down the ages a succession of prophets and great teachers have sought to convey to humanity the truths and the principles upon which we should base our lives and our conduct. Abraham, Buddha, Jesus, Mohamed, along with many other great seers and spiritual leaders. I sometimes try to imagine what might go through the heart and mind of Christ as He faces ands opens up to the human experience. If He is to return as some suggest, He cannot turn away. Perhaps His return can only occur through us. We are the mechanism of His facing up to the human situation. Or are we going to become His turning away? Will the Soul, the Christ within each of us, be allowed to find expression and, in a sense, provide the path for His return? The choice, I think, is ours.

Can we open our hearts to the collective human experience? Dare we open our hearts to the collective human experience? These are times that try our hearts and Souls to the limit, yes. That seems part of the deal as we move into the next millennium. Human hearts are awakening and it hurts. I always think of the rose bud, tearing itself apart so that it may unfold and reveal its hidden beauty for all to see. Many hearts remain tight, desperately trying not to burst under the pressure to open up. Yet as the rose bud must open in due season so must the human heart if we are to live to our potential as human beings. The closed heart enables the individual to remain insensitive to the plight and suffering of others. It remains closed to protect the individual from the hurt that will follow if they seek to reach out to another who is in pain.

We have to forget ourselves. We have to stand open to the flow of unconditional love that is the nature of the Soul and the energy of the heart. We have to find ways to cope with the hurt that it will bring us as we enter into truly heartfelt communion with our sisters and brothers everywhere. We have to love.

We surely live in the time of the call to love.

18

Epilogue

Throughout this book I have sought to highlight qualities that I think are important for the present and the future of humanity and evolution on our planet. I have set them in a diversity of contexts whilst endeavouring to show that there are connections running between them. In a way I have tried to illustrate that making sense of life is rather like putting together a jigsaw puzzle. I do not claim to have offered you all the pieces, I think this would be rather a grand claim for such a small volume as this. But I hope it will have triggered thought about which pieces you might need to look for in other places, and which ones you may already possess, but had not realised their significance.

I have not touched on some areas that many will consider important—what science is telling us, the environmental situation on Earth, perhaps I could have said more about what particular religions are telling us. That is not to in any way ignore or underestimate their importance. They are crucial. The environment is, perhaps, the one issue more than any other that has the capacity to force us all to face up to the consequences of our actions, choices and values that we live by. Yes, there are those who try to belittle what is happening, but that is usually linked to short-term economic interests—individual, organisational or national. Such attitudes can have no place in the future and are a constant block to the assembling of the jigsaw.

I mention science not just in terms of the environment but also in relation to the discoveries that are being made about the nature of the world around us, and in particular the striving for a greater understanding of consciousness and matter, and how they interact, with all the fascinating aspects of the sub-atomic and quantum worlds that provide the building blocks of our material world including the bodies that we each have. We are sure to see unexpected discoveries in these areas that may well challenge our world view of what reality actually is.

Finally, religion. Can core truths within religious traditions be agreed upon as to how we should live our lives and the ethical and moral codes that should govern our decisions and actions? In our often separative, selfish, consumer and profit driven world we desperately need some values to underpin codes of conduct.

I leave this chapter for you to complete, to reflect on what your jigsaw looks like, or maybe what you see as being part of the larger picture. You see, I cannot complete the jigsaw on my own. I can only complete my part, as you can only complete yours, but we need each others pieces to make the whole.

I wish you well with your striving to find the pieces that make your individual jigsaw of self, and in contributing to completing the jigsaw of life that is our planetary evolution. And who knows, if the jigsaw hypothesis is correct, we may once again be able to wonder if perhaps someone is observing the process. Perhaps, if we do find a positive and constuctive way forward, if we do manage to overcome the sense of separateness and self-interest that currently dominates the global landscape. If some semblance of harmony and beauty does become revealed and made manifest, then an Onlooker will begin to smile. Perhaps when this occurs it will be drawing near to the end of the seventh day and the Onlooker will see that it is Good. Perhaps on that day we, too, will know ourselves as we truly can become for we will have rediscovered the original goodness that is our basic human potential and our spiritual birthright.

I would like to end with a poem I wrote 10 years ago. I believe it captures something of the energy of our times and maybe, the hope and the promise that what we are doing, as we struggle with our individual and collective jigsaws, is part of a greater process that may attract the attention of … well, let us simply say the Onlooker that I have referred to previously. It is a poem that gives me hope when I read it, may it give you hope too.

The Way of Fire

Fire, blazing in the night,
Orange glow that drips delight,
Dancing fingers, tendrils flame,
Myriad sparks, not two the same.
Showering fire to pierce the gloom,
That holds us to a world of doom.
Burning through distorted thought,
The battles on, it must be fought.
Fiery pain in preparation;

End the world of separation.
The burning ground surrounds us all,
A distant voice, the spirit's call.

The fires rage, the water steams,
Nothing's real or what it seems.
Fogs obscure the hidden light;
Out of mind and out of sight.

Out the fire a beam breaks through,
An arrow which in silence flew,
Piercing veils and cutting clean,
A path of fire of wondrous sheen.
Speeding on, beyond the Sun,
Carrying the thoughts of everyone;
The hopes, appeals, the voiceless cry
Of suffering humanity.
It hits it's mark, response is swift,
Light and Love, a priceless gift,
Streams towards the wondrous jewel
Locked in combat, fate so cruel.

Fiery hearts and minds respond
To emanations from beyond
The world of sense in which we dwell,
Our self-created human hell.

A distant fire, burning bright,
So cold, so clear, comes into sight.
Form bound thought is now no more,
Humankind has reached the door.
Through pain and fiery agony
Is born the One Humanity.

A greater Glory now revealed,
No longer from our Souls concealed.
The fiery Will that knows no bound
Is heard in one Eternal Sound.

Books by the author

The 'Living Therapy' series, published by Radcliffe Publishing, Abingdon, UK:

Counselling for Problem Gambling
Counselling for Eating Disorders in Women
Counselling for Eating Disorders in Men
Counselling Young People
Relationship Counselling: Sons and their Mothers
Responding to a Serious Mental Health Problem
Counselling for Progressive Disability
Counselling a Recovering Drug User
Problem Drinking
Counselling a Survivor of Child Sexual Abuse
Counselling Victims of Warfare
Counselling for Obesity
Counselling Young Binge Drinkers
Counselling for Death and Dying
Time-limited Therapy in Primary Care
Person-centred Counselling Supervision
Workplace Counselling in the NHS

Models of Care for Drug Service Provision, Radcliffe Publishing, Abingdon, UK

Counselling the Person Beyond the Alcohol Problem, Jessica Kingsley Publishers, London, UK

A Little Book of Therapy, Pen Press Ltd, Brighton, UK

Binge!—a novel, iUniverse.com, USA & UK

The Sevenfold Circle: Self Awareness in Dance, Lynn Frances and Richard Bryant-Jefferies, Findhorn Press, Forres, Scotland

Index

978-0-595-48002-
0-595-48002-0

Printed in the United Kingdom
by Lightning Source UK Ltd.
126328UK00001B/151-225/A

9 780595 480029